T0335812

The (Pre-)Dawning of
Functional Specialization
in Physics

Frontispiece

In brief, functional specialties are functionally interdependent.

Such interdependence is of the greatest methodological interest. First, without any prejudice to unity, it divides and clarifies the process from data to results. Secondly, it provides an orderly link between field specialization, based on the division of data, and subject specialization, based on a classification of results. Thirdly, the unity of functional specialties will be found, I think, to overcome or, at least, counter-balance the endless divisions of field specialization.[*]

The need for some division is clear enough from the divisions that already exist and are recognized. ... First, then, the need is not simply a matter of convenience. ... It arises not to divide the same sort of task among many hands, but to distinguish different tasks and to prevent them from being confused. ... Secondly, there exist the different tasks. ... Thirdly, the distinction and division are needed to curb totalitarian ambitions. ... Fourthly, the distinction and division are needed to resist excessive demands.[†]

---

[*] Lonergan, B. (1975) *Method in Theology* (Darton, Longman & Todd, London), Functional Specialties, ch. 5, p. 126.

[†] Ibid, pp. 136–138.

The (Pre-)Dawning of Functional Specialization in Physics

Terrance J Quinn
Middle Tennessee State University, USA

World Scientific

NEW JERSEY · LONDON · SINGAPORE · BEIJING · SHANGHAI · HONG KONG · TAIPEI · CHENNAI · TOKYO

Published by

World Scientific Publishing Co. Pte. Ltd.

5 Toh Tuck Link, Singapore 596224

USA office: 27 Warren Street, Suite 401-402, Hackensack, NJ 07601

UK office: 57 Shelton Street, Covent Garden, London WC2H 9HE

Library of Congress Cataloging-in-Publication Data

Names: Quinn, Terrance J. (Professor of mathematics), author.

Title: The (pre-)dawning of functional specialization in physics /
    Terrance J. Quinn (Middle Tennessee State University, USA).

Description: Singapore ; Hackensack, NJ : World Scientific, [2017]

Identifiers: LCCN 2016054698| ISBN 9789813209091 (hardcover ; alk. paper) |
    ISBN 9813209097 (hardcover ; alk. paper)

Subjects: LCSH: Physics--Methodology. | Physics--Philosophy.

Classification: LCC QC6 .Q85 2017 | DDC 530.01--dc23

LC record available at https://lccn.loc.gov/2016054698

British Library Cataloguing-in-Publication Data

A catalogue record for this book is available from the British Library.

Copyright © 2017 by World Scientific Publishing Co. Pte. Ltd.

All rights reserved. This book, or parts thereof, may not be reproduced in any form or by any means, electronic or mechanical, including photocopying, recording or any information storage and retrieval system now known or to be invented, without written permission from the publisher.

For photocopying of material in this volume, please pay a copying fee through the Copyright Clearance Center, Inc., 222 Rosewood Drive, Danvers, MA 01923, USA. In this case permission to photocopy is not required from the publisher.

Typeset by Stallion Press

Email: enquiries@stallionpress.com

Printed in Singapore

For

Philip McShane

Scholar and Teacher

Contents

Preface

While the main focus of this book is physics, *functional specialization* will be omni-disciplinary.[1] The sciences are, of course, not all the same. However, within functional specialization, *disciplines* as presently conceived eventually will become obsolete.

What is functional specialization? In the Introduction, I give foreshadowings of a (pre-) emergent (normative) eightfold division of labor in the Academy. The chapters go into some detail, in preliminary description. The book, then, is introductory. A main purpose is to invite interest in the need and possibility of implementation.[2]

Functional specialization is not yet a mainstream topic, neither in physics nor elsewhere in the Academy. It was Bernard Lonergan who, in February 1965, made the major breakthrough, discovering the fundamental structuring of the new methodology[3]. In the late 1960's, however, Lonergan struggled with lung cancer[4]. Still, he managed to write about his new result. He did so first in a short dense article in 1969[5]. That article later became Chapter 5 of his book on the new methodology[6].

---

[1] This will emerge in the book.

[2] There are already a number of works in various areas that help bring out the need, nature and possibility of functional specialization. See, for example, Anderson [1996] (law), [2010] (economics); Benton [2010] (language studies); Brown and Duffy [2016] (a collection of essays from a number of areas and contexts); Drage [2005] (women studies); Henman [2011] (philosophy), [2016] (neuroscience); McNelis [2014] (housing science); McShane [1976], [1977], [1980] (musicology, linguistics, sciences), directly relevant to physics and the present book is McShane [2001], [2004], [2007a], [2007b]. There are also many works by Philip McShane, too numerous to list here in this footnote (see, http://www.philipmcshane.org/); Quinn [2003] (mathematics), [2012] (sciences), [2015] (general invitation); Shute [2013] (interpretative of Lonergan's breakthrough in economics).

[3] Lambert and McShane [2010], p. 160.

[4] Lambert and McShane [2010].

[5] Lonergan [1969].

[6] Lonergan [1975], ch. 5.

Neither Lonergan's paper nor his book, however, were the kind of expansion that, in better health, he might otherwise have produced[7].

But, the present book is not about Lonergan, at least, no more than an introduction to Chemistry is about Mendeleev. Where Mendeleev discovered a Periodic Table of chemical elements, verifiable in reactions, Lonergan discovered the possibility of a normative (functional) collaboration, verifiable in eight main tasks already (subtly but so far rather confusedly) present in the Academy. The transition to functional collaboration will, then, not lose what has already been attained. It will be a transition to a new control of meaning, a new economy of effort, a new effectiveness in the Academy. In particular, it will subsume and make more precise an already familiar two-fold division of labor between experimental and theoretical work in physics[8] (a division of labor also familiar in other sciences).

Is there really a need for development in methodology? Or, are we doing fine as we are? Without a doubt, physics, the other sciences, the arts and technologies all have been making remarkable progress over the last few centuries. And, for scholars who are busy working and surviving within the protective environment of a research institute or academic department, it may be difficult to see or sense the present-day crisis in education and scholarship. If, however, one looks further afield, it is obvious that we are struggling within patterns of cumulative crises that are global, cultural, political, economic and environmental. What is not so obvious is that the more visible crises are part of confusion that also pervades present-day science and education.

For some (slight) intimation of the fact that there are fundamental problems, we can look, for instance, to philosophy of science. Trends in modern philosophy of science include the effort to obtain results about "any possible world." As Smolin observed, that way

---

[7] Lambert and McShane, Epilogue, sec.2, "The Far Larger Work," p. 257. See also note 55 of ch. 2.

[8] "In physics we have come to accept a division of labor between theorists and experimentalists" (Jaffe and Quinn [1993]), p. 2.

philosophers were unable to discern a general strategy that would explain how science (in this world)[9] works. And the strategy that they did invent didn't bear much resemblance to what scientists actually do. The successful strategies were discovered over time and are embedded in the practices of the individual sciences.[10]

Other leading scholars have made similar observations. Michael Redhead noted a

deep seated misunderstanding between scientists and philosophers[11].

Redhead's remark is from a talk given at Cambridge in 1993, and Smolin's book was published in 2006. But, basic incompatibilities remain, between philosophy of science and what we might call field science. Certainly, this Preface is not the place to attempt detailed contribution to that discussion. Indeed, a message of the book is that effective resolution of the problem will need community functional collaboration[12]. But, by way of context, I can at least point to clues that, by looking to method, already are available.

Science is dynamic. Discoveries are made. Methods shift, change and develop. The whole scientific process is noticeably growing and on the move. However, prevailing methods in contemporary philosophy of science consist of debate about ideal structures, where both debate and structures are remote to scientific practice and experience. This is true also in the literature following from Kuhn's work.[13] Note, too, that the problem is not solved by merely moving the discussion to a more complex context of an emerging multiplicity of interdisciplinary collaborations[14].

---

[9] Parenthetic phrase mine.

[10] Smolin [2006], p. 298.

[11] Redhead [1995], p. 2.

[12] As will become increasingly evident within emerging functional collaboration, control of meaning will require self-attention within an implemented "generalized empirical method." See ch. 6 and Epilogue.

[13] Henman [2016], pp. 35-36.

[14] "Thus much of the unity of science debate, although it is the only widely known philosophical discussion that addresses scientific interdisciplinarity directly, has relatively little to offer for the explanation or the critical assessment of actual interdisciplinary exchanges" (Grüne-Yanoff and Mäki [2014], p. 53).

A different approach is possible. We may attend to scientific practice. And, by doing so, eight main tasks can be discerned. Not disagreeing with Smolin, but changing the last word of the above quotation - and that way subtly shifting emphasis - *the successful strategies have been and are being discovered over time and are embedded in the practices of the individual scientists.*[15]

Why is physics a main focus of the book? It is evident and now broadly acknowledged that physics is present in, and relevant to, all of science and philosophy. But, while there have already been publications on the need and possibility of functional specialization[16], there is not yet a book on functional specialization and physics.

This also leads me to the title of the present book. Readers in physics may well be familiar with Lochlainn O'Raifeartaigh's book, *The Dawning of Gauge Theory*. O'Raifeartaigh identified three stages in the development of gauge theory[17], and two phases.[18] Lonergan also identified three stages, but *stages of meaning* in human history[19]. And he spoke of

---

[15] The need and possibility of self-attention is somewhat revealed as we climb through the chapters of the book. This helps toward raising awareness that, in more advanced work, eventually we will need "generalized empirical method." See note 12.

[16] See note 2.

[17] A first stage was thanks to inspired contributions of Hermann Weyl. Weyl did not yet have clarity on the issue, but he "proposed that gauge invariance be elevated from the rank of a symmetry to that of a fundamental principle. ... The second stage consisted in generalizing the gauge invariance used in electromagnetism to a form that could be used for the nuclear interactions. ... The third stage consisted of the gradual realization of the fact that, contrary to first appearances, the Yang-Mills gauge theory, in a suitably modified form, was suitable for describing both of the nuclear interactions. None of the stages in the development of gauge theory were easy" (O'Raifeartaigh [1997], p. vii).

[18] "The emergence of gauge theory actually took place in two phases. In the first phase, which started with Einstein's gravitational theory and ended with Yang-Mills theory in the mid-fifties, the physicists were essentially making educated guesses, based on general principles such as symmetry, conservation laws and the need for a unified theory of gravitation and electromagnetism. ... The second phase, which lasted roughly from the mid-sixties to the mid-eighties, was concerned with the application of the existing theory to the nuclear forces, whose structure had by now become much clearer" O'Raifeartaigh [1997], p. 4).

[19] Lonergan [1972], Stages of Meaning, Section 3.10, pp. 85-99.

two times (of the temporal subject) in history[20]. Of course, O'Raifeartaigh and Lonergan were speaking about different things in different contexts. However, by playing off of the title of O'Raifeartaigh's book, the present title not only indicates that the book has a focus in physics, but also points to a key analogy. Like O'Raifeartaigh's book, the present book is about what can be called a gauge theory. But, rather than a mathematical gauge theory in physics, functional specialization will be a *practical gauge theory of human progress*.

In its maturity, functional specialization will be coincident with what Lonergan called the *third stage of meaning*[21]. Through implementation, the eightfold structuring of human progress will be gradually and increasingly revealed. As in the words of O'Raifeartaigh, none of the stages in the development are (or will be) easy[22]. Unlike contemporary gauge field theory, functional collaboration has not yet *dawned*. However, much as one might say in developmental biology, functional specialization is *pre*-emergent. For, even if their presence so far is shadowy and confused, neither yet differentiated nor coordinated, the eight types of activity, and combinations thereof, already are present and discernible. Identification and implementation will, among other things, provide a basis needed toward resolving questions both about 'interdisciplinary

---

[20] "Lonergan talks about the two times of the temporal subject. It is an evolutionary perspective of the emergence and development of humanity toward its own effective, collaborative, luminously self-appropriated living. There is a first lengthy time in which our thinking operates spontaneously; there is a second time in which a revelation of our human nature has occurred, our thinking is explicitly affirmed by us and our living is a differentiated self-directing. The first and third stages of meaning in *Method* (Lonergan [1972]) coincide with the two times of the temporal subject,..., while the second stage of meaning, simply put, is the (lengthy, axial) in-between time" (Gillis [2011], par. 2). See McShane [1998], Editor's Introduction, pp. 1–4. McShane has named the "in-between times" Axial Times.

[21] Lonergan [1972], pp. 93-99. Third stage of meaning: "a later time with at least a dominant authority of the mediation of *generalized empirical method*" (McShane [1998], p. 42). Italics mine. See note 12. Regarding stages of meaning, see also notes 19 and 20.

[22] See note 17.

research' as well for making progress in questions about science and philosophy of science.[23]

One of the intended groups of readers are scholars in the physics community who are interested in basic issues and fundamental directions. But, as mentioned above, functional specialization is not only for physics, and so the book also is intended, eventually, for readers in all areas of scholarship in the global Academic community - that is, readers who are interested in the possibility of a verifiable and implementable heuristics of progress. At this time in history, a difficulty for some readers will be that the book gradually brings out the need for self-attention[24]. Self-attention, though, is not yet standard practice, neither in science nor in philosophy of science. But, if one insists on familiar methods, then we can only expect familiar results. And, as history shows, some kind of development in method is desperately needed. Even if only at first in the context of this book, might it not be worth the experiment to find out whether or not attending to one's operations might shed some light on the problem?

Another group of intended readers are scholars in Lonergan Studies[25]. My hope is that this book might help scholars in Lonergan Studies appreciate something of the omni-disciplinary significance of Lonergan's discovery. I also hope that the book will help motivate scholars in Lonergan Studies to take up Lonergan's challenge to go beyond descriptive methods in philosophy, and so cultivate the (future) emergence of generalized empirical method (normatively grounded in scientific development).[26]

It is with pleasure that I include the following acknowledgements. I am very grateful to Middle Tennessee State University, for awarding me a Non-Instructional Assignment in the Fall semester of 2016, in order to finish this book. Thanks go to the anonymous referees for their constructive comments. I very much appreciate the time and energy given

---

[23] We are "going to have all the divisions coming from field specialization, all the divisions coming from subject specializations. Still, (we will) have a central core in which everything comes together" (Lonergan [2010], p. 455).

[24] See note 12.

[25] There are now numerous academic associations in the world that promote interest in Lonergan's work.

[26] See Quinn [2017] and references therein. Similarly, see McShane [1970].

by both John Benton and Meghan Allerton, for reading sections of the book and providing helpful comments. Allerton also provided me with some relevant background material in contemporary watershed science. Thanks to all of my colleagues who have been working on Lonergan's discovery and have been supportive of my efforts in this area. I would also like to thank the World Scientific Publishing Acquisitions Editor, Chad Hollingsworth (Hackensack, NJ); and the Desk Editor, Low Lerh Feng (Singapore). Chad helped make the initial proposal process go smoothly. Lerh Feng was a great help to me in the production of the final manuscript. Their assistance was much appreciated.

References

Anderson, B. (1996) *Discovery in Legal Decision-Making* (Kluwer Academic Publishers, Dortrecht).

Anderson, B. (2010) The Nine Lives of Legal Interpretation, *J. Macrodyn. Ana.*, vol. 5, pp. 30-36.

Benton, J. (2008) *Shaping the Future of Language Studies* (Axial Publishing, Canada).

Brown, P. and Duffy, J., eds. (2016) *Seeding Global Collaboration* (Axial Publishing, Vancouver).

Drage, A. (2005) *Thinking Woman* (Axial Publishing, Halifax, subsq. Vancouver).

Gillis, S. (2011) The Second Stage of Meaning, *Functional Specialization Seminars*, May 29, http://www.sgeme.org/BlogEngine/post/2011/05/29/The-Second-Stage-of-Meaning-by-Sandy-Gillis.aspx.

Grüne-Yanoff, T. and Mäki, U. (2014) Introduction: Interdisciplinary-Model Exchanges, *Stud. Hist. Phil. Sci.*, vol. 48, pp. 52-59.

Henman, R. (2011) An Ethics of Philosophic Work, *J. Macrodyn. Ana.*, vol. 7, pp. 44-53.

Henman, R. (2016) *Global Collaboration: Neuroscience as Paradigmatic* (Axial Publishing, Vancouver).

Jaffe, A. and Quinn, F. (1993) "Theoretical mathematics": Toward a cultural synthesis of mathematics and theoretical physics, *Bull. Amer. Math. Soc.* 29, pp. 1-13.

Lambert, P. and McShane, P. (2010) *Bernard Lonergan. His Life and Leading Ideas* (Axial Publishing, Vancouver).

Lonergan, B. (1969) Functional Specialties in Theology, *Gregorianum*, vol. 50, pp. 485-505.

Lonergan, B. (1972/73/75) *Method in Theology* (Darton, Longman & Todd, London; rep., (1990), University of Toronto Press, Toronto)

Lonergan, B. (2010) *Early Works on Theological Method 1*, eds. Doran, R. M. and Croken, R. C., vol. 22 in *Collected Works of Bernard Lonergan* (University of Toronto Press, Toronto).

McNelis, S. (2014) *Making Progress in Housing: A Framework for Collaborative Research* (Routledge, Oxford).

McShane, P. (1970) *Randomness, Statistics and Emergence* (University of Notre Dame Press, Notre Dame, IN).

McShane, P. (1976) *Shaping the Foundations* (University Press of America, Lanham, MD).

McShane, P. (1977) *Economics for Everyone - Das Jus Kapital* (Edmonton: Commonwealth Press, Edmonton; subsq., Axial Publishing, Vancouver).

McShane, P. (1980) *Lonergan's Challenge to the University and the Economy* (University Press of America, Lanham, MD).

McShane, P. (1998) *A Brief History of Tongue. From Big Bang to Coloured Wholes* (Axial Publishing, Halifax, subsq. Vancouver).

McShane, P. (2001), Elevating *Insight*: Space-Time as Paradigm Problem, *Method: J. Lonergan Studies*, vol. 19, no. 2, pp. 203 – 229.

McShane, P. (2004) The Origins and Goals of Functional Specialization, http://www.philipmcshane.ca/quod-17.pdf.

McShane, P. (2007a) *Method in Theology: Revisions and Implementations*, online book: http://www.philipmcshane.org/website-books/.

McShane, P. (2007b) *Lonergan's Standard Model of Effective Global Inquiry*, online book: http://www.philipmcshane.org/website-books/.

Quinn, T. (2003) Reflections on Progress in Mathematics, *J. Macrodyn. Ana.*, vol. 3, pp. 97-116.

Quinn, T. (2012) Invitation to Functional Collaboration: Dynamics of Progress in the Sciences, Technologies and Arts, *J. Macrodyn. Ana.*, vol. 7, pp. 92-120.

Quinn, T. (2015) Community Climbing: Toward Functional Collaboration, *J. Macrodyn. Anal.*, vol 8, pp. 45–66.

Quinn, T. (2017) *Invitation to Generalized Empirical Method* (World Scientific Publishing, Singapore).

Redhead, M. (1995) *From Physics to Metaphysics. The Tarner Lectures, Cambridge, 1993* (Cambridge University Press, Cambridge).

Smolin, L. (2006) *The Trouble with Physics, The Rise of String Theory, the Fall of a Science, and What Comes Next* (Houghton Mifflin, Boston).

Shute, M. (2013) Functional Collaboration as the Implementation of 'Lonergan's *Method*' Part 1: For What Problem is Functional Collaboration the Solution? *Divyadaan: Journal of Philosophy and Education*, vol. 24, No. 1, pp. 159-190. Also (2015), *J. Macrodyn. Ana.*, vol. 8, pp. 93–116.

Introduction

Foreshadowings of a Normative Division of Labor

The story of physics certainly includes Archimedes'[1] discovery of the principle of displacement, and his subsequent books on hydrostatics[2]. About nineteen centuries later, Galileo's[3] inquiries into free-fall led not only to the discovery of a new law, but also to a fundamental development in method[4]. There is little doubt that Archimedes appealed to experience and, in particular, applied his result to find out that the King's crown was, as suspected, not pure gold. But, as Stillman Drake observed, prior to Galileo,

> the systematic appeal to experience in support of mathematical laws seems to have been lacking. ... The design of experiments to discover new mathematical laws comes after Galileo's time.[5]

That shift in method led to centuries of major development in physics and, of course, continues to be a main feature of modern physics. In fact, also included in that major development were further developments in method:

> In physics we have come to accept a division of labor between theorists and experimentalists. But in fact only recently has the division become so clear. Until

---

[1] Archimedes: c. 287 B.C.E – c. 212 B.C.E

[2] Heath [2002], On Floating Bodies, Book I, pp. 253 – 262; and Book II, pp. 263–300.

[3] Galileo: 1564 C.E. – 1642 C. E.

[4] Drake [1970], p. 44.

[5] Eventually, Kepler's method and discoveries also contributed to these later developments in method.

the beginning of the twentieth century there was basically one community of physicists. It was the ideal, and by and large the practice, that the same people both speculated about theory and verified their speculations with laboratory experience. Certainly in Europe it had become clear by 1900 that a bifurcation had occurred: there were sufficiently many physicists who concentrated solely on the theoretical side of their work that one could identify two distinct communities. This trend proceeded somewhat more slowly in the United States. E. C. Kemble, who worked at Harvard in the area of quantum theory, is generally regarded as the first American to obtain a doctorate in purely theoretical physics (though his 1917 thesis contained an experimental appendix). ... Distinctness of the theoretical and experimental physics communities should not be confused with their independence. Theory (has been) vital for experimentalists to identify crucial tests and to interpret the data. Experiment (has been) vital for theorists to correct and to guide their speculations.[6]

In 20th century physics, an often dramatic back-and-forth between developing experimental and theoretical physics eventually led up to what is now known as the Standard Model[7]. And, at present, even while there are open problems[8], the Standard Model is still being verified at the highest energy levels produced in the Large Hadron Collider (CERN)[9].

---

[6] Jaffe and Quinn [1993], pp. 2–3. In the last sentence of the quotation, in parentheses, I changed the verb-conjugation from 'is' to 'has been.' Jaffe and Quinn seem to mean both. The adjustment fits the past-oriented sentences earlier in the quotation.

[7] Some details of that development are discussed in ch. 3, Functional History.

[8] Open problems include: hierarchy; the number of parameters; recent puzzles about dark energy and dark matter; and the anomalous magnetic moment of the muon. Note that, while there have been numerous attempts (quantum loop gravity, etc.), there is still no consensus on how to relate general relativity and the Standard Model. For more detailed listings of open problems, see, e.g., a slightly dated 2012 list, gathered together by John Baez. For instance, from the section called Particle Physics, two questions are: 11: "Why do the particles have the precise masses they do? Or is this an unanswerable question?"; and 12: "Why are the strengths of the fundamental forces (electromagnetism, weak and strong forces, and gravity) what they are? For example, why is the fine structure constant that measures the strength of electromagnetism, about $1/137.036$? Where do such dimensionless constants come from? Or is this an unanswerable question?" (Baez [2012]). See also, Smolin [2006], The Five Great Problems in Theoretical Physics, Part I, ch. 1, pp. 3–17.

[9] As of August 5, 2016, in experiments at CERN: "No significant excess is observed over the standard model predictions" (Search for resonant production of high mass photon pairs using 12.9fb^{-1} of proton-proton collisions at $\sqrt{s} = 13$ TeV and combined interpretation of searches at 8 and 13 TeV, *CMS Physics Analysis Summaries, CMS Collaboration*, Particle Physics Experiment, http://cds.cern.ch/record/2205245).

There are, however, influential subgroups in present-day Theoretical Physics who hope to replace the Standard Model with supersymmetric models and (super-) string theories. The idea would be to

supplant the standard understanding of particle physics, solv(e) deep problems in physics and (explain) the nature of the universe's mysterious dark matter.[10]

That there are difficulties involved with these aspirations is well-known.[11] My purpose in this book, though, is not to contribute directly to debate about whether or not supersymmetric models or string theories might replace the present Standard Model. The community will eventually sort that out. The main focus of the book is method, and so I wish to draw attention to part of what has been driving that debate. In advocating for string theories and supersymmetry, a common justification is the elegance of the mathematics involved. On the other hand, many leading physicists[12] are concerned because, among other things, neither is the landscape of (order 10^{500}) string theories verifiable by observation nor is there any data available to support the existence of supersymmetric particles.

Why, then, is there still debate about the matter? In fact, there are fundamentally different views about the role of theoretical physics in physics. Two main groupings of views are summarized as follows:

One (view) is that the role of theory is to be closely tied to the experimental and phenomenological field, help experimentalists interpret their experiments and distinguish signals from noise. The other attitude is that the goal of theoretical physics is to achieve a higher level of understanding. In order to achieve such

---

[10] Kykken and Spiropulu [2014], p. 36. Verb tenses adjusted to fit context.

[11] "What is string theory? We really do not understand what string theory is at its core" (Gross [2005], p. 5903). As of September 2016, there is no evidence for particles predicted by supersymmetric models. The basic hypotheses of string theory imply a landscape of more than 10^{500} versions (or, if hypotheses about the "background manifold" are relaxed, infinitely many versions), none of which are explicitly known. "String theory has moved from being a theory that explains everything to a theory where almost anything is possible" (Ellis [2011], p. 42). The "unshakable fidelity to supersymmetry is widely shared. Particle theorists do admit, however, that the idea of natural supersymmetry is already in trouble and headed for the dustbin of history unless superpartners are discovered soon" (Kykken and Spiropulu [2014], p. 36).

[12] See, for example, Smolin [2006].

understanding one might focus on the solution of well-defined mathematical models consistent with general physical principles, regardless of whether those models are real of not. Indeed, how much value do we assign to simplicity and mathematical elegance? That is what the second group usually cares about. [13]

Now, notice here that the summary just quoted is neither experimental physics nor theoretical physics. And, yet, is the statement by Gross not a contribution to physics? In other words, there is thought and scholarship in physics that is neither experimental physics nor theoretical physics. In the example, Gross edges into history and philosophy of physics. It is true that Gross is not first known for his work in history or philosophy of physics. For, only a subgroup of the community formally publishes in the history or philosophy of physics. However, some scholars in physics challenge whether or not historical understanding and philosophy of science are properly part of, and influential in, experimental and theoretical physics.

Yet, if one argues to the contrary and pursues one's interests accordingly, does that not (self-) reveal a performance-contradiction? And so, Rovelli was able to make an observation about

> scientists who say 'I don't care about philosophy' – it's not true that they don't care about philosophy, because they *have*[14] a philosophy. They're using a philosophy of science. They're applying a methodology. ... They're taking a position. [15]

As further elementary self-attention reveals, whether or not spelled out within a professional context, one's grasp of developments (such as from Newton to Einstein and from Hermann Weyl to Ronald Shaw), one's ideas of what physics is about and what physics is for, all are part of one's view, part of what shapes, directs and guides one's questions, insights, and plans for ongoing collaborations and applications. And, as is abundantly clear from ongoing controversies (supersymmetric theories and string theories versus what is sometimes called *traditional* physics – where *traditional*

---

[13] Gross [2005], p. 5909. For his own view, Gross suggests a combination: "(B)oth approaches are good, and both are necessary. Each is strengthened by the other. They are both part of what I take to be a theoretical physicist" (Gross [2005], p. 5909).
[14] Italics in original text.
[15] Rovelli [2014], p. 228.

physics asks for data[16]), differences in views about physics have been, and are, leading to differences in directions in experimental work, theoretical work, collaborations, applications and education.

There are, then, different camps. Moreover, the number of specialized topics continues to increase[17]. And, because

(p)hysics is becoming so big, with so many different areas, ... some (have) asked: 'Will Physics split into different fields, different disciplines?'[18]

Whether or not there is to be a "split into different fields, different disciplines," a basic question remains: What is physics?

David Gross describes something of the challenge:

The great tradition of physics is to preserve a core of common education and culture that defines a physicist, even as physicists expand their horizons and take advantage of opportunities in neighboring fields. In fact, the unity of physics was demonstrated by the success of (the) anniversary conference[19], where we successfully brought

---

[16] See, e.g., note 60 in ch. 3.

[17] In broad outline, some of these include: field theories and particle physics; thermodynamics; physics of fluids and complexity; and cosmology. There is quantum (physics of) chemistry; and biophysics (which includes physics of contemporary neuroscience, and consciousness), as well as physics of exotic materials, molecular machines, as well as chemical and biomolecular materials. The physics of geography is part of contemporary environmental science. Radio carbon-dating is used in archeology and art history. Developments in new technologies emerge from and contribute to advances in experimental methods in physics and other sciences. The engineering sciences depend on, and often contribute to physics in, for example, the ongoing development of technologies for the manufacturing of clothing, furniture and household goods; musical instruments and other technologies for the arts, water systems, roads, bridges, buildings, vehicles of all kinds, computers, cell phones and other communications technologies, nanotechnology, and modern weaponry. There are ongoing developments in physics for modern medicine (Jeraj [2009]). There is scholarship in physics education and psychology. There is ethics of physics in energy, society and economics (Smil [2011] and Amaldi [2005]). Also included are historical studies and philosophy of physics. Theology, too, now has recognized linkages with contemporary physics (see, for example, Mann [2014] and Newsome [2011], although this will not be a topic in this book).

[18] Gross [2005], p. 5908.

[19] *The Future of Physics, A celebration of 25 Years of The Kavli Institute for Theoretical Physics*, October 7–9, 2004.

together world leaders in all subfields of physics, from cosmology to biophysics, to discuss the future of physics as an intellectual and cultural enterprise.[20]

Eventually, though, we need precision about these issues. For instance: What is a "horizon" and what, in particular, is one's own "horizon"? What is the actual "unity of physics"? What is a "neighboring field" or "subfield"? How is the "future of physics" determined? What is the human intellect, and what is an "intellectual and cultural enterprise"?

This Introduction began by drawing attention to a now-familiar two-fold division of labor that, through the pressures of history, emerged in the early 20th century: experimental physics and theoretical physics. From our few observations here, it is also obvious that experimental physics and theoretical physics are only part of physics. What else is there? What else has been, and is influencing the ongoing global physics community?

With Lonergan's discovery[21] in mind, it turns out to be fruitful to ask about main tasks in physics. It is an empirical question. That is, besides experimental and theoretical physics, we can seek to identify what other tasks are present in the field. And so, the first eight chapters of this book are to help reveal the operative presence of, in fact, *eight* main tasks internal to physics. The names for these tasks are not essential. But, as introduced by Lonergan, they are *functional specialties*: *functional research*; *interpretation*; *history*; and *dialectics*; and *functional foundations*; *doctrines* or *policies*; *systematics*; and *communications*.[22]

Here, then, is a brief outline of the book.

Chapter 1, Functional Research, is on the work of *finding relevant data, written or otherwise*[23]. Experimental physics is a familiar context. But, the chapter is to help us see that obtaining data through experiment is only one part of a much broader task.

---

[20] Gross [2005], p. 5908.

[21] See Preface.

[22] Lonergan [1969]; Lonergan [1975], ch. 5, pp. 125–144.

[23] Expressions in italics are quotations taken from a listing of the eight functional specialties given in McShane [1998], p. 106. Italics here make for a smoother presentation.

Chapter 2, Functional Interpretation, is on the work of attempting to *reach the meaning of those who produced data*[24]. This task includes some of what is now called theoretical physics, at least, when the effort is to explain data obtained from experiment and observation. But, what is data? Even in experimental physics, results are communicated by researchers using digital imaging technologies, symbols, charts and tables of measurements, in the language of physics and within the modern context of a Standard Model. And so, data potentially relevant to progress includes writings of authors, as well as reported events and occurrences in world communities. But, that means that the challenge of (functional) interpretation intrinsically depends on implementing up-to-date methods of hermeneutics in the scientific and historic context.

The title of Chapter 3, Functional History, may seem self-explanatory: the chapter looks to the task of historical understanding. However, by attending to examples (e.g., Lochlainn O'Raifeartaigh's book on the emergence of gauge theory[25]), we find evidence of (an almost-emergent) historical analysis, the task of *connecting meanings of writings and doings, etc.*

Written histories, though, are many. Not only are topics investigated diverse, but historians work from diverse points of view. Nevertheless, some scholars go on to reach for holistic views, seek to identify the best and worst of what has been and is going on, *seek to identify the best basic directions*. There is, then, evidence of a fourth task in physics. Chapter 4 is about that fourth task, namely, Functional Dialectics.

The first four chapters are on tasks where scholars contribute to progress in physics, but do so within a focus on past or present events and occurrences. But, the physics community, and indeed, the entire Academic community, also looks to the future. Chapter 5, Functional Foundations, is on the first main task in that thinking toward the future. In the scientific

---

[24] Note that the word 'interpretation' is being used here in the sense of reaching the meanings of expressions of individuals. This is not to be confused with the use of the word in philosophy of physics, where we find such expressions as 'interpretation of quantum mechanics.' The two usages are not unrelated and can be sorted out later as we go through the book.

[25] O'Raifeartaigh [1997].

context, Chapter 5 is about the task of *expressing the best fundamental (in the sense that they are not tied to age, time, etc.) directions.*

There is a further need of *reaching relevant basic pragmatic truths* and values. Chapter 6, Functional Doctrines, is on the task of working these out in the layerings of an up-to-date scientific-philosophic context, in the light of functional foundations.

For a moment, I defer comment on Chapter 7, to say something about Chapter 8. Scholars communicate with both scholars and non-scholars, with those whose main work is academic as well as with those whose main work is non-academic. Theoretical physicists converse with historians; and with those who work on up-to-date heuristics; but also with engineers, people in industry, lawyers, politicians, economists and educators. Foundations in physics are discussed with theologians. And so on. There are communications of all kinds. While not yet often adverted to, latent in such communications is understanding and selection of what needs to be communicated, as well as how to communicate with particular groups or audiences. Already present in the community, then, is the need and possibility of a practical science of communications, dialogue with other views that is *local collaborative reflection.* In this book, that task is called Functional Communications.

But, what is selected and what is communicated? Normally, communications are intended to sustain and promote progress (whatever one means by progress). What are the possibilities for progress, not ideal types (such as what dominates discussion in contemporary philosophy of science), but really possible in given situations? Prior to functional communications, then, there is the task of *drawing correctly and contra-factually on the strategies and discoveries of the past to envisage time-ordered possibilities.* Results here need to be up-to-date and compatible with fundamental directions, and be explanatory of basic pragmatic truths and values. And so, Chapter 7 is on Functional Systematics, a seventh task that (self-evidently) is subsequent to doctrines but prior to communications.

The first eight chapters of the book partly are to help us make some progress toward noticing and describing eight distinct tasks. However, the significance of Lonergan's discovery is not only that these tasks are present and operative. Recall that while the division of labor between experimental and theoretical physics is now familiar, it was not always so. Lonergan discovered that there are not only two, but eight distinct tasks. So far, however, these eight tasks tend to be only obscurely present[26]. The preliminary efficiency of the two-fold division of labor will be seen to be (pre-functional) precursor to what eventually will be an effective eightfold division of labor.

Where each of the first eight chapters of the book include modest beginnings in envisioning a few aspects of each task, the tasks are functionally related. And so, there is the need for developing heuristics for the entire collaboration, all functional specialties working together. Chapter 9 is a flex in that direction. The chapter provides descriptions and diagrams for various aspects of global functional collaboration. The emergence and differentiation of functional collaboration will include collaborations and communications of the form C_{ij}, $i, j = 1, 2, 3, ..., 8$. But, the Academic enterprise is, of course, only one part of world cultures, arts, technologies, education and economies. Within the ethos of functional collaboration, world process will include collaborations and communications C_{99} in a *common plane of meaning*[27], as well as C_{i9}, $i = 1, 2, ..., 8$. Explanation adds to experience and so experimental data can be said to be communicated *forward* to theoretical physics. But, historical understanding includes connecting (some) prior theories. In that sense, the results of theoretical physics also are moved *forward*, for they provide material for historical understanding. In a similar way, but within a future functional control of meaning, there will be a central flow that will be forward and cumulative: C_{91}, C_{12}, C_{23}, ..., C_{78}, C_{89}. Communications of the form C_{19} are the Academy attending to the world; and C_{89} are where functional collaboration bears fruit in the world. Economics will be a sub-structuring of the new collaboration. It is a major issue and, as it happens, the basic field equations also were discovered by Lonergan. I am pointing here to fundamental future developments. Chapter 9, though, is brief, consisting mainly of diagrams, each with some commentary. This is because, while most of the preliminary heuristics and diagrams in this chapter are to express what is relatively invariant, specifics of that future

---

[26] For an example and discussion, see Appendix A.
[27] Lambert and McShane [2010], The Tower of Able: Lonergan's Dream, p. 163.

collaboration will need to be worked out as we go, through implementation. As Aristotle observed:

For the things we have to learn before we can do them, we learn by doing them.[28]

Both prior to and subsequent to the possibility of an eightfold division of labor is a basic problem of control of meaning in the sciences and philosophy of science. Progress toward functional collaboration will undoubtedly accelerate progress in control of meaning, to be obtained through "generalized empirical method."[29] The Epilogue also points to various issues mentioned only in passing in the book. Some references are given to supplementary and supporting literature. The Epilogue ends by inviting a choice regarding functional collaboration.

There is also an Appendix. The Appendix looks to two samples from the literature in, respectively, the history of physics and philosophy of science. The samples are from a book by Max Jammer[30]; and from an article by Alexander Bird[31]. The discussion is not meant to be either conclusive or complete. It is to get our feet wet with an elementary exercise. The purpose of the exercise is two-fold: (1) to make beginnings in being able to identify the presence of different tasks in contemporary literature in history and philosophy of science; and (2) to make beginnings in noticing that, at this stage of history, that presence in the literature tends to be confused and not adverted to, thereby undermining effectiveness. The aim is not to single out these two authors as such. Nor are the reflections intended as a serious contribution to interpretation of their works. At the same time, their methods are representative of present-day standards. To learn more about the presence of the eight distinct tasks, more complete and increasingly precise exercises will be needed – for the works of Jammer and Bird, and for work throughout the Academy. Once functional collaboration becomes operative, the works of Jammer and Bird

---

[28] Ross, tr. [2001], Book II, par. 1.
[29] The need and possibility of such a method emerges gradually in the book. Explicit definitions are given ch. 6. See Preface, note 26.
[30] Jammer [2000]. Max Jammer (b. 1915 – d. 2010).
[31] Bird [2007].

will contribute to progress through functional re-cycling. The same comments apply to the potentially significant works of past and present authors in all areas.[32]

References

Amaldi, U. (2006) *The New Physics for the Twenty-First Century*, ed. Fraser, G., Chapter 19, Physics and Society (Cambridge University Press, Cambridge), pp. 505-531.

Baez, J. (2012) *Open Questions in Physics* (http://math.ucr.edu/home/baez/physics/).

Bird, A. (2007) What is Scientific Progress? *Noûs*, 41:1, pp. 92–117.

Drake, S. (1970) *Galileo Studies* (The University of Michigan Press, Ann Arbor).

Ellis, G. F. R. (2011) Does the Multiverse Really Exist? *Sci. Am.*, Aug., pp. 39–43.

Gross, D. (2005), The Future of Physics, *Int. J. Mod. Phys. A,* vol. 20, no. 26, pp. 5897–5909.

Heath, T. L. (2002) *The Works of Archimedes* (Dover Publications, Mineola, NY; 1st pub., (1897), Cambridge University Press)

Jaffe, A. and Quinn, F. (1993) "Theoretical mathematics": Toward a cultural synthesis of mathematics and theoretical physics, *Bull. Amer. Math. Soc.* 29, pp. 1–13.

Jeraj, R. (2009) Future of Physics in Medicine and Biology, *Acta Oncol.*, 48, pp. 178–184.

Jammer, M. (2000) *Concepts of Mass in Contemporary Physics and Philosophy* (Princeton University Press, Princeton, NJ).

Kykken, J. and Spiropulu, M (2014) Sypersymmetry and the Crisis in Physics, *Sci. Am.*, May, pp. 34–39.

Lonergan, B. (1969) Functional Specialties in Theology, *Gregorianum*, vol. 50, pp. 485–505.

Lambert, P. and McShane, P. (2010) *Bernard Lonergan. His Life and Leading Ideas* (Axial Publishing, Vancouver).

Lonergan, B. (1975/73/72) *Method in Theology* (Darton, Longman & Todd, London).

Mann, R. B. (2014) Physics at the Theological Frontiers, *PSCF*, vol. 66, no. 1, pp. 1–12.

McShane, P. (1998) *A Brief History of Tongue. From Big Bang to Coloured Wholes* (Axial Publishing, 1st pr., Halifax, subsq. Vancouver).

Newsome, W. T. (2011) Life of science, life of faith, ch. 36 in Chiao, R. Y. et al, eds., *Visions of Discovery. New Light on Physics, Cosmology, and Consciousness* (Cambridge University Press, Cambridge), pp. 730–750.

O'Raifeartaigh, L. (1997) *The Dawning of Gauge Theory* (Princeton University Press, Princeton).

Ross, W. D. tr. (2001) *The Nicomachean Ethics* (Virginia Tech, Blacksburg).

---

[32] See also references in note 2 of Preface.

Rovelli, C. (2014), Science Is Not About Certainty (with an Introduction by Lee Smolin), ch. 12 in Brockman, J. ed., *The Universe, Leading Scientists Explore the Origin, Mysteries, and Future of the Cosmos* (New York: Harper Perennial, 2014), pp. 214–228.

Smil, V. (2011) Science, energy, ethics, and civilization, ch. 35 in Chiao, R. Y. et al, eds., *Visions of Discovery. New Light on Physics, Cosmology, and Consciousness* (Cambridge University Press, Cambridge), pp. 709–729.

Smolin, L. (2006) *The Trouble with Physics, The Rise of String Theory, the Fall of a Science, and What Comes Next* (Houghton Mifflin, Boston).

Chapter 1

Functional Research

1.1 Introduction: A Pilot-Wave

Over the last few centuries, techniques and technologies in experimental physics have developed greatly. Contemporary cyclotrons are among the most advanced machineries in the world; and there are modern astronomical observatories such as the James Ax Observatory in Chile. If, however, we look to all of physics, in addition to experiments and astronomical observations, there are also many other sources of data that contribute to progress in physics.

To begin exploring this further, we can, for instance, look to fluid dynamics, where there has been a renewed interest in de Broglie's theory of pilot-waves[1]. In the article by Oza *et al*[2], the authors provide new computational results for the

rich nonlinear dynamics that arise when a droplet walks in a rotating frame.[3]

In addition to the numerical part of their results, also in their paper is the message that de Broglie's theory of pilot-waves may contribute to new results in other areas of physics.

---

[1] Oza at al [2014].

[2] "Since the walking drop is propelled through resonant interaction with its own wave field, it represents the first macroscopic realization of a double-wave pilot-wave system, a theoretical framework for a rational quantum mechanics first envisioned by Louis de Broglie. We here explore the rich nonlinear dynamics that arise when a droplet walks in a rotating frame" (Oza at al., [2014], p. 082101-1).

[3] Oza et al., [2014], p. 082101-1

As first demonstrated in Yves Couder's laboratory, these droplets exhibit several phenomena previously thought to be exclusive to the microscopic quantum realm. Specifically, the walking-drop system exhibits analogs of single- and double-slit diffraction, tunneling, orbital quantization, level-splitting, and wavelike statistics in confined geometries. Since the walking drop is propelled through resonant interaction with its own wave field, it represents the first macroscopic realization of a double-wave pilot-wave system, a theoretical framework for a rational quantum mechanics first envisioned by Louis de Broglie.[4]

John Bush, one of the co-authors of the article just cited, gives additional context:

> The key question is whether a real quantum dynamics, of the general form suggested by de Broglie and the walking drops, might underlie quantum statistics. While undoubtedly complex, it would replace the philosophical vagaries of quantum mechanics with a concrete dynamical theory. ... de Broglie's pilot-wave theory deserves a second look.[5]

Why does de Broglie's pilot-wave theory deserve a second look? Oza *et al* provide evidence that making use of the old theory can: provide new results in computational fluid dynamics; may lead to new insight into quantum mechanics and, by extension, would have implications for quantum field theory, and the philosophy of physics; moreover, in a subsequent review article, Bush points out that the pilot-wave theory probably also will have applications in stochastic electrodynamics[6].

To what extent Oza *et al* are on to something important is for the community to determine. The focus of this book is methodology. And so, I invite the reader to notice something that is implicit in the clustering of works just cited. Let's look again to Bush's remark: *de Broglie's pilot-wave theory deserves a second look*[7]. The pilot-wave theory caught the attention of these authors. To their minds, the de Broglie pilot-wave theory

---

[4] Oza et al., [2014], p. 082101-1.
[5] Hardesty [2014].
[6] Bush [2015].
[7] Bush [2015].

stands out, in new light relative to their present-day (2014) physics. They went on to give evidence for how the theory might advance contemporary computational fluid dynamics, as well as how it might contribute to progress in several other areas of physics. But, distinct from that follow-up work is an insight. Implicitly, at least, there is an initial insight that calls for a better understanding of de Broglie's pilot-wave theory.

How might de Broglie's pilot-wave theory help advance contemporary physics? Might the theory eventually contribute to the emergence of a new Standard Model? Will it, perhaps, be helpful in reaching a new understanding of the present Standard Model? Will it play into progress in philosophy of physics? Might the pilot-wave theory be helpful in the development of new technologies in society? While Oza *et al* follow up with various computational results, the insight that de Broglie's work deserves new inquiry is, of course, open ended. We don't yet know how significant the theory eventually might be. But, there is an insight, namely, that de Broglie's theory (possibly) deserves a second look.

1.2 Directed Research

The immediate context of works cited in Section 1.1 is fluid dynamics. But, noticing and then communicating that de Broglie's pilot-wave theory might warrant further inquiry implicitly reveals a kind of work that occurs not just in fluid dynamics, but throughout the sciences. For, in all areas, there is the work of being alert to, hunting for and communicating potentially significant data.

What is potentially significant data? I am not attempting to provide a formal definition here. I am using the expression 'potentially significant data' descriptively and non-technically. In that sense, data is considered to be potentially significant when researchers think that it might contribute to progress. Of course, a familiar illustration is found in experimental physics, where researchers expertly and creatively are on the look-out for such data. But, while potentially significant data often is obtained through experiment and observation, as Bush's comment helps reveal[8], potentially significant data also can be something that some author has written.

---

[8] See note 5.

I have given only a loose indication of what I mean by the name 'potentially significant data.' Finding data that is potentially significant; and also reaching a better understanding of that task, will need the collaboration of future scholars. We can, though, broaden our sense of the problem, by looking to a few more examples.

In the LHC at CERN, billions of collisions per second can be recorded. However, not all particle-tracks are of particular interest to research teams. Data that intrigue, that attract attention, are data that seem to be in some way anomalous[9].

There are famous examples from history. In 1859, based on an analysis of data from almost two hundred years of observations, Urbain Le Verrier reported that the precession of the perihelion of the planet Mercury was not accounted for by Newtonian theory[10]. In 1881 there was the Michelson-Morley experiment, a negative result about previously expected measurable motion of objects through a supposed "luminiferous aether."[11]

As pointed to in Section 1.1, there are other sources of data that contribute to progress in physics, besides experimental physics. There was, for example, the anomaly that was Einstein's 1905 paper in which he introduced special relativity. Or, for the reader familiar with the history of gauge invariance, one might think of Weyl's 1919 paper (his unsuccessful first attempt to unify gravity and electromagnetism). Weyl's paper did not fit with prevailing views but did come under intense scrutiny.[12]

For physicists with sufficient background, it was evident that neither the works of Einstein nor of Weyl could fit with standard views of their

---

[9] The use of the word *anomaly* here is descriptive and elementary, 'something that does not fit.' See also note 13.

[10] Le Verrier [1859].

[11] Michelson [1881].

[12] Several lead physicists of the time went on to the further task of attempting to understand Weyl's paper. The reception was mixed. It is now known that his "brilliant proposal contains the germs of all mathematical aspects of non-Abelian gauge theory" (Straumann [2005]). See also O'Raifeartaigh and Straumann [2000]; and O'Raifeartaigh [1997], for more details on how Weyl's work originally was received by the community. Regarding 'interpretation,' see note 17.

times. But, both articles eventually led to major shifts in physics[13]. To end this section, I quote Bernard Lonergan. In the quotation (below), Lonergan was speaking about teamwork within a functional division of tasks[14]. His comments apply, in particular, to the task of finding data, and of sharing those findings with the academic community.

(Y)ou can have teamwork insofar, first of all, as the fact of reciprocal dependence is understood and appreciated. Not only is that understanding required; one has to be familiar also with what is called the *acquis*, what has been settled, what no one has any doubt about at the present time. You're doing a big thing when you can upset that, but you have to know where things stand at the present time, what has already been achieved, to be able to see what is new in its novelty as a consequence.[15]

1.3 Sources of Data

In Section 1.2, there was not a double-slit experiment, but a double-observation: (1) Already operative in physics is the task of finding potentially significant data; and (2) while familiar examples of that task are found in experimental physics, that task looks to all data, including data that are words of authors.

The purpose of Section 1.3 is to begin to bring out a few subtleties of data research. To do that, let's look to a few more examples.

Appealing to what (at the time) were puzzling numerical results obtained from computer models, Edward Lorenz (1917–2008)[16] broke through to seminal work on chaotic weather patterns. In turn, new

---

[13] This is preliminary description. It is not intended as classification, such as in Kuhn's work, where scientific revolution is idealized as being preceded by an accumulation of experimental anomalies. However, as Smolin pointed out, what Kuhn observed "describes what has happened in some cases" (Smolin [2006], p. 115).

[14] Lonergan [2010], Method (continued), Functional Specialties, and an Introduction to Horizons and Categories, pp. 441–472.

[15] Lonergan [2010], p. 462.

[16] Chang [2008]. See also, Branner [2008], pp. 495–496.

interpretations[17] of his writings have been part of ongoing advances in both micro and macro thermodynamics[18] and fluid dynamics[19].

Since Keeling's first observations in 1958[20], measurements of CO_2 concentrations in the earth's atmosphere have led to the emergence and ongoing development of what now is a multidisciplinary climatology[21] that involves all of the major sciences, technologies, economics, and more.

The writings of John Wheeler also are data, and by some in the contemporary physics community are considered to be potentially significant for future progress:

> The symposium *Science and Ultimate Reality* held in Princeton, March 15-18, 2002, sought to honor John Wheeler in his ninetieth birthday year and to celebrate his sweeping vision of the physical universe and humankind's place within it. In keeping with Wheeler's far-sightedness, the symposium dwelt less on retrospection and more on carrying Wheeler's vision into a new century.[22]

There are also anomalies in our experience in contemporary physics. For, how are we to reconcile every-day descriptive understanding of Space and Time with space-time determined by families of elementary particles mutually defined within Lie algebras, and with statistical distributions in (rather) exotic function spaces (Standard Model, Supersymmetry, String theory, and so on)? Or, how are we to reconcile our every-day experience of Space and Time with data such as obtained by CERN, which mainly

---

[17] This use of the word 'interpretation' foreshadows ch. 2. It is not be confused with the word as used in contemporary philosophy of physics, such as in discussions about various 'interpretations of quantum mechanics.' In later chapters, the context of discussion will include philosophy of physics.

[18] Ross and Berry [2008].

[19] Oza et al [2014].

[20] Keeling et al. [1976].

[21] "The climate system is one of the most complex subject(s) of study that one can imagine: multiscale, multi-process, featuring a gigantic range of dynamics that are only partially and indirectly observed. Besides the problematic of global warming, studying the climate system is also an unprecedented scientific adventure bringing together scientists form different disciplines in an era strongly influenced by the paradigms of computer simulation" (Crucifix, Lucarini and Vannitsem [2014]).

[22] Davies [2004], p. 10.

consist of computer-screen images and numerical and graphical data relative to convenient scales? Is technical data and mathematical understanding of such data in terms of a Standard Model ultimately some kind of (perhaps temporary) substitute for what elementary particles really look like, if only we could get a better look[23]? Making use of Wheeler's words,

(w)e have to move from the imposing structure of science over into the foundation of elementary acts of observer-participancy.[24]

Data from experiments have their special significance within the scientific enterprise. But, it is also generally acknowledged in contemporary foundations of physics and human biophysics, and is verifiable in one's own performance, that all data are within human consciousness[25]. And so, even in apparently straightforward cases (like reporting on numerical results from an experiment), whatever else one's results might mean, one is expressing something about one's experience. Moreover, such communication is done by inviting further complex layerings of experiences in oneself and others – through diagrams, graphs, words, and symbols within our sensitive consciousness, as well as other "acts of observer-participancy" [26] (such as understandings and decisions).

---

[23] Debates are ongoing. See, e.g., Rosen, S. M. [2013]. In fact, the problem was solved by Lonergan: "a mistaken twist was given to scientific method at the Renaissance" (Lonergan [1992], p. 107). Through self-attention, one can witness the death of the mistaken conflict between so-called *primary qualities* and *secondary qualities*, introduced into philosophy of science by Galileo. See Section 1.7, of *Invitation to Generalized Empirical Method*, Quinn [in press].

[24] See note 22, p. 23. See also Wheeler and Zurek [1983], p. 210.

[25] We are touching here on advanced problems that go beyond the scope of this book. For precise descriptive results, see Lonergan [2002], The Nature of Consciousness, pp. 157–169. Within generalized empirical method, modern accounts of consciousness will need to include advances in neuroscience, psychology.

[26] Regarding "acts of observer-participancy," there are contributions by scholars who, at present, are perhaps less familiar to readers from the physics community. For that reason, I put the following discussion in footnote. First, there is the quotation from Wheeler, which continues: "No one who has lived through the revolutions made in our time by relativity and quantum mechanics .... can doubt the power of theoretical physics to grapple with this

still greater challenge. ... Recent decades have taught us that ... (the) scope of physics is immensely greater than we once realized. We are no longer satisfied with insights only into particles, or fields of force, or geometry, or even space and time. Today we demand of physics some understanding of existence itself" (Wheeler and Zurek [1983], p. 210; see also, Davies [2004], p. 23). Wheeler's insights probably eventually will be recycled, once physics matures to become an eightfold cyclic functional collaboration. In the meantime, there is a philosophical tradition that also has grappled with the problem of "observer-participancy," but in different ways. I am thinking here, in particular, of the work of Merleau-Ponty, and then also Renaud Barbaras. In *Phenomenology of Perception*, Merleau-Ponty writes: "Our bodily experience of movement is not a particular case of knowledge; it provides us with a way of access to the world and the object, with a 'praktognosia,' which has to be recognized as original and perhaps as originary" (quoted in, Barbaras [1991], p. xxiii). The now famous book by Barbaras goes on to a complex weave of results - partly historical, partly dialectical, but mainly (descriptive) interpretation of the life-work of Merleau-Ponty: "As for the interpretation of the evolution of Merleau-Ponty's thought as a whole, I believe that I can maintain my developmental hypothesis starting from the problem of ideality and truth, that is, from the necessity of passing from a phenomenology of perception – open to the reproach of being nothing other than a psychology of perception – to a *philosophy* of perception, discovering in perception a mode of being that holds good for every possible being" (Barbaras [2004], p. xxi). "Our intention is neither to take up the movement of (Merleau-Ponty's) thought from one end to the other in order to grasp *The Invisible and the Invisible* (Merleau-Ponty [1968]) as its end point, nor to evaluate his heritage as developed elsewhere. Our intention is to reconceive Merleau-Ponty's ontology on its own terms – which is, in our eyes, the key to this entire thought as well as to numerous contemporary reflections" (Barbaras [1991], p. xxxi). However, within this first chapter of the present book, that to which I wish to draw attention is not Barbaras' interpretation of Merleau-Ponty's work, nor the complex weave of historical or dialectical elements in *The Being of the Phenomenon*, but, rather, that aspect of Barbaras' work which is precursor to *functional research* (see second last paragraph of Section 1.4). For that, we may look partly to the "Preface to the English Translation" (Barbaras [1991], pp. xix–xxiv), but mainly to the "Introduction to the French Edition" (Barbaras [1991], pp. xxvii–xxxiv). In the "Preface to the English Translation," Barbaras provides indirect evidence for reasons to study Merleau-Ponty's work, work "whose full import we have not yet measured" (Barbaras [1991], p. xx). In the original "Introduction," however, Barbaras is more direct. He highlights various details of the work of Merleau-Ponty, and indicates ways in which the work of Merleau-Ponty would seem to be anomalous within the philosophical tradition. Barbaras then also discusses the potential importance of Merleau-Ponty's work – to development in philosophy generally, as well as for philosophy of science. One of Barbaras' summary comments is: "To us, it looks as though we have here at our disposal material whose richness justifies the attempt to explicate this never-completed work. ... We could characterize the spirit in which we have

1.4 Anomalies in Contemporary Collaboration

Experimental and theoretical physics emerged gradually, through the pressures of history. As physics made progress over the last few centuries, the need to distinguish the two tasks became more or less unavoidable, and obviously has been immensely practical. This division of labor not only shapes scholars' careers, but generally is preserved within articles, books and other publications. Reports from CERN, for example, typically concentrate on potentially significant data; while reports from the Theoretical Physics division of *Nature*, say, focus on "models for understanding empirical results or constructing self-logical theories for explain phenomena beyond current experiments."[27]

Controversies from string theory aside[28], an ongoing division of labor between experimental and theoretical physics is, *de facto*, operative. However, observations from previous sections of this chapter help reveal that the task of seeking potentially significant data is not limited to data obtained through experimental physics. The task, self-evidently, has no prescribed generic or disciplinary boundaries[29]. This presents us with various difficulties. But, this also hints at the possibility of a new differentiation within the physics community, indeed, within the scientific community.

Why do I suggest that there are difficulties? Data research is familiar in experimental physics and, as four centuries of progress shows, seems to present us with no special difficulties. But, implicitly, the task is operative throughout all of physics, including experimental and theoretical physics. We have, then, a basic anomaly in contemporary method: The task is

---

considered these texts by saying that we have read them as if they were *classics*" (Barbaras [1991], p. xxxi, italics in original text). The "Introduction to the French Edition" is a contribution toward identifying data that might contribute to progress in the field. Note that methods of self-attention and reflection practiced and advocated by Merleau-Ponty, and then further developed in Barbaras [2006], are not mainstream in contemporary philosophy of physics. In that sense, it may be noted that my inviting attention to their works in the context of the present article is a further hinting of the possibility of functional research.

[27] http://www.nature.com/subjects/theoretical-physics.

[28] See, for example, Smolin [2006].

[29] See note 33.

verifiably present and operative throughout all of physics, but in areas of inquiry that do not explicitly focus on data from *experimental* physics, the task generally is not adverted to[30].

Are we not being nudged here by history? Might it not be to our advantage to bring an already operative task into the open? Such an effort would gradually reduce attempts to inadvertently combine data research with other tasks such as interpretation and foundational reflection[31]. We also can expect a new efficiency. In experimental physics and theoretical physics, differentiation of data research is now more or less necessary and practical. How much more necessary and practical will it be to attain differentiation of data research in, for example, interpretation of authors' works, as well as other more complex (and multidisciplinary) inquiries?[32] The name *functional research* is for that future development, for data research explicitly differentiated, identified, implemented and developing within physics and the Academy.

How far might we go in such a development in method? At this stage, we only have hintings of a future possibility. Also, major stumbling blocks may come to mind[33]. Still, imagine, if you will, something like the mind-

---

[30] See note 31.

[31] At this point, I intend this as a question to keep in mind. That *ad hoc* combinations of task is a problem in contemporary scholarship will become more evident toward the end of the book. Appendix A looks at two examples from the literature: in history of physics; and philosophy of physics. Discussion of each is preliminary and superficial, but helps reveal that *ad hoc* transitioning between main tasks undermines the possibility of contributing effectively to progress.

[32] See note 31.

[33] The data research task has no generic, specific or disciplinary boundaries. How are particular data to be combined or discussed relative to what even now is an omnidisciplinary *acquis*? For now, this is just a question, but a question that is a foreshadowing of a needed and possible heuristics of layerings being discovered through physics; chemistry; and so on. As mentioned in the Preface, reaching such a heuristics will need a generalized empirical method, pointings to which can be found in works cited there. But, will a differentiated data research therefore require that a data researcher be expert in all areas? As already implicit in the contemporary division of labor between experimental physics and the rest of physics, differentiation of the task will relieve the individual of excessive demands in at least two main ways: (1) a data researcher need not simultaneously attempt data research, interpretation of authors' works, history, and so on; and (2) scholars in data research can be expected to have areas of particular interest and expertise.

set and proficiency of contemporary experimental physics, but (eventually) a differentiated omnidisciplinary mind-set and proficiency. Much as in contemporary experimental physics, different scholars will have different areas of expertise. But, as a task, functional research will look not only to potentially significant data available through experiment, but will be open to all sources of data in the global and historical physics community[34].

1.5 References

Barbaras, R. (2004) *The Being of the Phenomenon, Merleau-Ponty's Ontology*, tr., Ted Toadvine and Leonard Lawlor (Indiana University Press, Bloomington and Indianapolis). Orig. French version, (1991) (Éditions Jérôme Millon).

Barbaras, R. (2006) *Desire and Distance, Introduction to the Phenomenology of Perception*, tr. Milan, P. B. (Stanford University Press, Stanford).

Branner, B. (2008) Dynamics, sec. 1.5 of article IV.14, in Growers, T., Barrow-Green, J. and Leader, I., eds., *The Princeton Companion to Mathematics* (Princeton University Press, Princeton), pp. 495–496.

Chang, K. (2008), Edward N. Lorenz, a Meteorologist and a Father of Chaos Theory, Dies at 90, *New York Times*, April 17 (2008).

Davies, P. C. W. (2004), John Archibald Wheeler and the clash of ideas, ch. 1 in Barrow, J. D., Davies, P.C.W. and Harper, C.L, *Science and Ultimate Reality* (Cambridge University Press, Cambridge), pp. 3–26.

Crucifix, M., Lucarini, V. and Vannitsem, S. (2014) Advances in Climate Theory, Brussels (Uccle) 25–27 August 2014, Brussels-Belgium, http://www.climate.be/advances/.

Hardesty, L. (2014), Fluid mechanics suggests alternative to quantum orthodoxy. New math explains dynamics of fluid systems that mimic many peculiarities of quantum mechanics, *MIT News Office*, September 12, http://newsoffice.mit.edu/2014/fluid-systems-quantum-mechanics-091.

Keeling, C. D., Piper, S. C., Bollenbacher, A. F. and Walker, J.S. (1976), Atmospheric carbon dioxide variations at Mauna Loa Observatory, Hawaii, *Tellus*, Vol. 28 (1976), pp. 538–551.

Le Verrier, U. (1859) (in French), Letter de M. Le Verrier à M. Faye sur la théorie de Mercure et sur la mouvement du périhélie de cette planète, *Comptes rendus*

---

[34] As outlined in the Introduction, the other seven main tasks will be discussed in subsequent chapters. Once operative, all main tasks also will be sources of potentially significant data.

hebdomoadaires des séances de l'Académie des sciences (Paris), vol. 49, pp. 379–383.

Lonergan, B. (1992) *Insight: A Study of Human Understanding*, vol. 3 in Crowe, F. E. and Doran, R. M., *Collected Works of Bernard Lonergan* (University of Toronto Press, Toronto).

Lonergan, B. (2002) *The Ontological and Psychological Constitution of Christ*, tr. by Michael G. Shields, Vol. 7 of *Collected Works of Bernard Lonergan* (University of Toronto Press, Toronto).

Lonergan, B. (2010), *Early Works on Theological Method 1*, eds., Doran, R. M. and Croken, R. C., Vol. 22 of *Collected Works of Bernard Lonergan* (University of Toronto Press, Toronto).

Merleau-Ponty, M. (1968) *The Visible and the Invisible, Followed by Working Notes*, tr. by Lingis, A., ed. Claude Lefort (Northwestern University Press, Evanston, IL). Orig. French version, (1964) *Le Visible et l'invisible* (Editions Gallimard, Paris).

Michelson, A. A. (1881), The Relative Motion of the Earth and the Luminiferous Aether. *Amer. J. Sci.* 22, pp. 120–129, 1881.

O'Raifeartaigh, L. (1997) *The Dawning of Gauge Theory* (Princeton University Press, Princeton).

O'Raifeartaigh, L, and Straumann, N. (2000), Gauge theory: historical origins and some modern developments, *Rev. Mod. Phys.*, Vol. 72, no. 1, pp. 1–23.

Oza, A. U., Wind-Willassen, Ø., Harris, D.M., Rosales, R. R. and Bush, J.W. M. (2014), Pilot-wave hydrodynamics in a rotating frame: Exotic orbits, *Phys. Fluids*, 26, pp. 082101-1–082101-16.

Quinn, T. (2017), *Invitation to Generalized Empirical Method* (World Scientific Publishing, Singapore).

Rosen, S. M. (2013) Bridging the "Two Cultures": Merleau-Ponty and the Crisis in Modern Physics, *Cosmos and Hist: J. Nat. Soc. Phil.*, vol. 9, no. 2, pp. 1–12.

Ross, J. and Berry, S. R. (2008), *Thermodynamics and Fluctuations far from Equilibrium*, *Springer Series in Chemical Physics*, vol. 90 (Springer, New York).

Smolin, L. (2006), *The Trouble with Physics. The Rise of String Theory, the Fall of a Science, and What Comes Next* (Houghton-Mifflin, Boston).

Straumann, N. (2005) Gauge Principle and QED, Invited talk at *PHOTON2005, The Photon: Its First Hundred Years and the Future*, 31.8 - 04.09, Warsaw, https://arxiv.org/pdf/hep-ph/0509116.pdf.

Wheeler, J. A. and Zurek, W. H. (1983) *Quantum Theory and Measurement*, Princeton Legacy Library (Princeton University Press, Princeton).

Chapter 2

Functional Interpretation

2.1 Introduction

No doubt, experimental data is not the same as data that are expressions of authors. However, even experimental data are known through expressions of authors and so, in both cases, the question arises: What is the meaning of whoever produced the data? In other words, there is the problem of interpretation[1]. Evidently, interpretation of data is a task that follows the task of obtaining potentially significant data. Part of what is needed, though, is progress toward (self-) coherent and implementable heuristics of this distinct task, here called *functional interpretation*.

Following the approach described in the Introduction[2], this chapter includes descriptions of two efforts in interpretation. Section 2.2 is on O'Raifeartaigh's interpretation of Weyl's gauge theory. Sometimes the positive work of influential authors also carries elements of confusion. Section 2.3 looks to McShane's interpretative comments on Schrödinger's book *Space-Time Structure*[3]. McShane, of course, recognizes the importance of Schrödinger's work but also invites us to note certain anomalies that call for attention. Drawing on the results of Sections 2.2 and 2.3, Section 2.4 provides hintings of the high bar to be met in functional interpretation. Also revealed in Sections 2.2 and 2.3 is that functional interpretation has historical dimensions. For instance, even though there is no evidence that Weyl foresaw later applications of

---

[1] See Introduction, note 22.
[2] See paragraph in Introduction that immediately precedes the outline of the book.
[3] Schrödinger [1950].

25

his ideas (in physics or, indeed, in, e.g., technologies of modern medicine), the fact is that

> (h)istorically, ..., Weyl's 1929 papers were a watershed. They enshrined as fundamental the modern principle of gauge invariance, in which the existence of the 4-vector potentials (and field strengths) follow from the requirement that the matter equations be invariant under gauge transformations ... of the matter fields. This principle is the touchstone of the theory of gauge fields, so dominant in theoretical physics in the second half of the 20th century.[4]

Section 2.5 provides a few comments regarding interpretation and history. All along I am taking help from Lonergan's pointers. The purpose of Section 2.6 is twofold: (1) Lonergan's work on interpretation is referenced, indicating the possibility of major developments in interpretation; and (2) it also becomes evident that it will be quite some time before we can expect functional interpretation of Lonergan's results on interpretation.

2.2   O'Raifeartaigh Interpreting Weyl's Gauge Theory

I start here by looking to interpretative work of Lochlainn O'Raifeartaigh (1933–2000) which is part of his book *The Dawning of Gauge Theory*[5]. In Chapter 1 of the book, O'Raifeartaigh includes: a translation of the publication of Hermann Weyl's first (unsuccessful) attempt (1918) to unify gravitation and electricity[6]; *Postscript* remarks by Einstein on Weyl's paper, and Weyl's *Author's reply* to Einstein[7].

In Chapter 5 of his book, O'Raifeartaigh includes a translation of Weyl's later (successful) 1929 "Classic." [8] At the beginning of that chapter, O'Raifeartaigh writes:

> From the philosophical point of view, the (1929) paper marked the completion of his 1918 ideas. He had always been convinced that there was a close analogy

---

[4] Jackson and Okun [2001], p. 676.

[5] O'Raifeartaigh [1997].

[6] O'Raifeartaigh [1997], "Gravitation and Electricity by H. Weyl (1918)," ch. 1, pp. 13–23.

[7] O'Raifeartaigh [1997], pp. 24–37.

[8] O'Raifeartaigh [1997], "Weyl's Classic, 1929," ch. 5, pp. 107–144.

between gravitation and electromagnetism and was particularly impressed by the resemblance between the derivations of charge conservation and energy-conservation in the respective theories. In this paper he was able to formulate the analogies between the two theories explicitly by means of the tetrad formalism and was able to overcome the objection to his 1918 paper, by adopting London's reinterpretation of the non-integrable scale factor of the metric as a non-integrable phase factor of the wave function[9].

In that same Chapter 5, O'Raifeartaigh goes on to provide a detailed account of each section of Weyl's 1929 paper, because

each one is worth some comment.[10]

Of course, one needs to read O'Raifeartaigh's book to appreciate his detailed discussion of Weyl's 1929 paper. And, in fact, O'Raifeartaigh's book includes elements of both interpretation and history (and more)[11]. But, in particular, we see O'Raifeartaigh expressing a refined understanding of a paper written by Weyl in 1929.

2.3 McShane Interpreting Schrödinger's *Space-Time Structure*[12]

O'Raifeartaigh's book mainly emphasizes positive aspects of Weyl's gauge theory[13]. To learn something more about the task of interpretation, it can help to look to an example where work being interpreted partly helped advance the field, but also has been problematic in some of its content and influence.

I draw attention, then, to interpretative comments given by Philip McShane, regarding Erwin Schrödinger's book *Space-Time Structure*[14]. Actually, the main focus of McShane's article[15] is not interpretation of

---

[9] O'Raifeartaigh [1997], p. 107.
[10] O'Raifeartaigh [1997], p. 109. The descriptions of Weyl's paper are on pp. 107–120.
[11] See chapters 3 and 4.
[12] Schrödinger [1950].
[13] See, however, O'Raifeartaigh [1997], pp. 34–36.
[14] Schrödinger [1950].
[15] McShane [2001].

Schrödinger's writings[16]. However, within his article, McShane includes what, for our present purposes, is an enlightening foray into Schrödinger's *Space-Time Structure*[17].

I return then to Schroedinger's work[18], the basic assumption of which is Einstein's view that the dynamic interaction of electro-gravitational realities grounds an intrinsic geometric structure of space-time. It is worth quoting Schroedinger at some length here: later we will see how it expresses compactly the central problematic. "In fact, though not always in wording, the mystic concept of force is wholly abandoned. Any 'agent'" whatsoever, producing ostensible accelerations, does so qua amounting to an energy-momentum tensor and via the gravitational field connected with the latter. The case of 'pure gravitational interaction' is distinguished only by being the simplest of its kind, inasmuch as the energy-momentum- (or matter-) tensor can here be regarded as located in minute specks of matter (the particles or mass-points) and as having a particularly simple form, while, for example, an electrically charged particle is connected with a matter-tensor spread throughout the space around it and of a rather complicated form even when the particle is at rest. This has, of course, the consequence that in such a case we are in patent need of field-laws for the matter-tensor (for example, for the electromagnetic field), laws that one would also like to conceive as purely geometrical restrictions on the structure of space-time" (Erwin Schroedinger, *Space-Time Structure* (Cambridge, Cambridge University Press, (1950), 1-2). ...

Note that *agent* is in quotation marks, warranted perhaps by the physicist's unease at intimations of efficient causality. At all events it is immediately cloudily avoided by the introduction of *particles*, merely gravitation or also charged, as *sources* of the energy-tensor. ... Putting one element very simply, Schroedinger is not in control of his use of nouns, ... As he moves into chapter one he immediately seeks "mathematical entities" that would ground a labeling, "a list of (grammatical) subjects without predicates" (Schroedinger, 7). Already words like "individuality" and "property" haunt the text, but he pushes on to the security of identifying familiar tensors of rank 1: "the nature of the entity may be such that there is a quadruplet of numbers attached to every point, varying from point to point" (Schroedinger, 16).

---

[16] "(T)his essay will reach toward showing that physics indeed can help us along the way to a richer appreciation of the strategy of functional specialization that is the main topic of *Method in Theology*. So I would focus your attention now on the word "paradigm." Other zones of interest could do a parallel job, and my hope is that this effort will encourage parallel efforts that should lead to the full elevation that is my topic" (McShane [2001], p. 203).

[17] Schrödinger [1950].

[18] McShane is referring to Schrödinger [1950].

Immediately, then, a need is evident. Certainly, one can say that the use of the word "entity" is inconsequential: nouns are a dime a dozen. Still, "things" call in quite another coinage. Schroedinger's book moves forward comfortably with entities called "tensors" and "densities," "more component entities" (Schroedinger, 16) and his discussion is complexified by connections and derivatives yielding formulae "easily memorized in the face of a bewildering dance of indices" (Schroedinger, 32). Do the indices relate to a spread of properties? So one pushes on to search for curvature tensors and geodesics, which finally brings us to the recurrence–seven times–of the word "particle" on page 57 (Schroedinger, 57). But the particles on that page are not regarded as agents or sources: "We assume that a gravitational field can be pictured as a purely geometrical property of space-time, namely as an affinity imposed upon it, and that it amounts to a geometrical constraint on particles. This affine connexion is to be regarded as an inherent property of the space-time continuum, not as something that is created only when there is a gravitational field" (Schroedinger, 58).

The need mentioned is now, perhaps, evident? It is the need for a redirection and elevation, that would prevent a space-time continuum from becoming a hypothetical thing with properties.[19]

If you read McShane's article, you will notice an unusual feature, namely, the suggestion that there is a need for us to read Schrödinger with a

new control of meaning to be grounded in the luminous character of writer, reader, and thinker.[20]

If one takes McShane's lead here, then, with modest beginnings in self-attention, it can indeed soon become evident and self-evident that

Schroedinger is not in control of his use of nouns.[21]

Yet, is this just nitpicking over Schrödinger's grammar? It would seem not. As noted in McShane's article, if one enters into a line by line reading

---

[19] McShane [2001], pp. 205–207.

[20] McShane [2001], p. 206. In the present book, while implicit from the beginning, the need of self-attention is developed only gradually. See, in particular, Introduction, chapters 4, 6, 7 and Epilogue. The Epilogue invites readers to a return to elementary exercises in self-attention. Such exercises will be needed prior to reaching a control of meaning in the contemporary sciences.

[21] See second paragraph of quotation from McShane, note 19.

of Schrödinger's book, identifying that lack of control highlights a confusion that weaves through the entire text. And, being aware of that confusion can help

> prevent a space-time continuum from becoming a hypothetical thing with properties[22].

At this time in history, many readers will be puzzled by, or perhaps dismiss as metaphor, McShane's claim that

> (t)he deeper issue is the fostering of a new control of meaning to be grounded in the luminous character of writer, reader, and thinker.[23]

But, is one to exclude McShane's suggestion out of hand, in particular, as it implies needed development in methods of interpretation? As history reveals, methods have developed gradually through ongoing collaboration. However, I am not attempting, here, to contribute to that larger effort. It is, though, a thesis of this book that methods will develop better once the community begins making progress toward an eightfold functional collaboration. For the present context, I merely offer indirect evidence for the need of at least some kind of development in methods in interpretation. As indicated below, that indirect evidence also is contextualized within this book.

Whatever one's epistemology, the fact is that traditional methods in interpretation, history and philosophy of physics are sustaining an ongoing multiplication of variously compatible and incompatible views about, for instance, quantum mechanics and general relativity. In particular,

> the foundations of quantum mechanics … remain hotly debated in the scientific community, and no consensus on essential questions has been reached.[24]

Some have gone so far as to claim that there is a crisis[25] in contemporary physics, or at least that physics is in a peculiar situation[26].

---

[22] McShane [2001], see last paragraph of note 19.
[23] McShane [2001], p. 206.
[24] Schlosshauer, Kofler and Zeilinger [2013], p. 222.
[25] Kykhen and Spiropulu [2014]; and Smolin [2013].
[26] Schlosshauer, Kofler and Zeilinger [2013], p. 230.

Even though evidence here is indirect, might not this sustained lack of consensus confirm that some kind of development in method might be desirable? But the question of how to promote and sustain ongoing development in methods is a central topic of the present book. And, the viability of an eightfold collaboration will be better judged at the end of the book.

One also might be nudged toward self-attention by noticing that in asserting that one's view is different from McShane's, one's focus is no longer Schrödinger's book. One is taking issue, instead, with another scholar's basic view. But, in that observation and self-observation, do we not then have further data on scientific method? For, such differences are normal. And, in asserting a difference in one's basic views, there is, then, the (foggy) presence of a task that evidently and self-evidently is distinct from interpretation of, say, Schrödinger's work. In other words, we obtain evidence for a task that (in its maturity) will, among other things, call for mutual encounter over differences (and similarities) in basic views and fundamental directions. At this stage of history, however, that task generally is not adverted to. It will become explicit and normative with and within the emergence of a fourth functional specialty that Lonergan called *functional dialectics*, to be discussed in more detail in Chapter 4.

2.4 Advanced Content and Analogy

O'Raifeartaigh draws attention to the fact that Weyl partly was inspired by

> analogy between gravitation and electromagnetism and was particularly impressed by the resemblance between the derivations of charge conservation and energy-conservation in the respective theories.[27]

McShane looks to

---

[27] See note 9. For some of Weyl's own comments, see O'Raifeartaigh [1997], pp. 32 (Weyl, 1918) and 37 (Weyl, 1955).

Schroedinger's work, the basic assumption of which is Einstein's view that the dynamic interaction of electro-gravitational realities grounds an intrinsic geometric structure of space-time.[28]

As is evident in both cases, the author (Weyl or Schrödinger) may (or may not) have attained a control of meaning appropriate to the context of their work. But, if the interpreter is to speak about what the author wrote in physics, undoubtedly, the interpreter needs to have attained an understanding and control of meaning that embraces the author's context in physics.

We get some inkling, then, that the context and control of meaning needed by an interpreter will be advanced indeed, when authors whose work is being investigated are lead thinkers like Weyl and Schrödinger. And so, for example, there is a (not-yet-present) advanced control of meaning needed when authors whose meaning is sought use words such as: analogy, derivation, charge-conservation, energy-conservation, theories, metric, non-integrable, phase factor, wave function (Weyl); and agent, energy-momentum tensor, gravitational field, matter, particles, electrically charged particle, field-laws, electromagnetic field, structure of space-time (Schrödinger); and so on.

This helps bring out still further aspects of interpretation. In the history and philosophy literature, titles such as "Bohm on dialogue," "Husserl on phenomenology," "Conversation between author A and author B," and so on, are standard. But, are not such titles (and typical claims therein, such as "Husserl meant …"; or "According to Lonergan, …, *followed by words of an interpreter*") often rather misleading? For, while a scholar might say "Husserl meant such and such," obviously, a scholar does not in fact have the same thoughts as Husserl. Unless there is a transcript provided of a conversation between authors A and B[29], calling a scholarly article on two authors "A Conversation between A and B" is, of course, just metaphor.

---

[28] See note 15.

[29] There was, for example, an "intense exchange of letters between Einstein and Weyl, part of which has now been published in Vol. 8 of *The Collected Papers of Albert Einstein* (1987)" (O'Raifeartaigh and Straumann [2000], p. 1).

This is not to suggest that the task of interpretation does not include attempting to work out what an author meant or might say.[30] But, it only contributes to ongoing confusion when what an interpreter says and means is assumed to be equivalent to what an author said and meant, "if only the author had done a better job at expressing themselves."

What I am touching on here is a principle of interpretation, a principle that may be obvious to many, but is not generally adverted to. An interpreter works within their own horizon[31]. An interpreter can express their own understanding; and give their own reasons for why they think that their understanding of what an author wrote probably is in some way related to the author's understanding. Interpretation, then, partly is based in analogy, analogy between what an interpreter understands and what an interpreter purports to be an understanding reached by an author whose meaning is being interpreted.[32]

2.5 Interpretation and History

Let's continue making use O'Raifeartaigh's book as well as McShane's comments about Schrödinger's book, to see what else we can learn about interpretation.

Beginning with O'Raifeartaigh, his direct comments on Weyl's gauge theory are only one part of his larger interest, namely, *The Dawning of Gauge Theory*[33] in the 20th century. He sheds light on how "Weyl's Classic, 1929"[34] paper was a turning point in Weyl's own understanding.

---

[30] This is a complex and subtle issue for future interpretation. See Lonergan's pointers, in Lonergan [1992], Section 17.3.6, The Sketch, pp. 602–603. In particular, see "hypothetical expression," p. 602. See also Section 2.6.

[31] "(W)hat lies beyond one's horizon is simply outside the range of one's knowledge and interests: one neither knows nor cares. But what lies within one's horizon is in some measure, great or small, an object of interest and of knowledge" (Lonergan [1975], p. 236). See also ch. 5, notes 14 and 16.

[32] The challenge of interpretation becomes especially evident when the work being investigated is written by a genius–like Einstein, Schrödinger, Weyl, Noether, Feynman, et al, and in methodology, Lonergan. On the challenge of interpreting Lonergan's work, see Section 2.6.

[33] O'Raifeartaigh [1997].

[34] O'Raifeartaigh [1997], ch. 5, pp. 107–144.

But, in his description of Weyl's work he does much more. Weyl's article emerged in an historical context. O'Raifeartaigh brings out various aspects of that historical significance. He identifies key moments in the historical development of gauge theory, with a deliberate focus on the first phase of its development that otherwise

is perhaps less well-known[35].

In fact, by the end of the book, O'Raifeartaigh's interpretative work sheds light on all three stages of what, as it happened, was a

slow and tortuous process that took more than sixty years.[36]

And so, in the concluding remark of his book, O'Raifeartaigh is able to say that

(a)ll major advances of recent years, such as grand unification, supersymmetry and string theory, have been made within this framework.[37]

I will come back to O'Raifeartaigh's work in a moment. But, let's now look again to McShane's remarks about Schrödinger's *Space-Time Structure*. Schrödinger's book is brilliant, but, as mentioned in Section 2.3, it also conveys a notion of Space-Time as some kind of

hypothetical thing with properties.[38]

One may or may not agree that this is a problem. But, the notion evidently is present throughout the literature of the second half of the 20th century, and so far also is part of the literature of the 21st century. Schrödinger's work is, in fact, part of a tradition that includes, for instance, Einstein's work (of course), the work of Hawking and Ellis on the *large scale structure of space-time*[39], recent work on gravitational "ripples in

---

[35] O'Raifeartaigh [1997], p. 4.
[36] O'Raifeartaigh [1997], p. vii.
[37] O'Raifeartaigh [1997], p. 242.
[38] McShane [2001], p. 207.
[39] Hawking and Ellis [1973].

spacetime," and so on. In contemporary science journalism, too, we see ideas with that same family trait:

> Many researchers believe that physics will not be complete until it can explain not just the behavior of space and time, but where these entities come from.[40]

Much as in O'Raifeartaigh's book vis-à-vis understanding Weyl's contributions in history, working out the meaning of Schrödinger's work also needs to include progress in identifying origins and influences in history. Especially significant are works whose results explicitly or implicitly draw on, or involve, ideas similar to those expressed in Schrödinger's works. Then, as O'Raifeartaigh was able to say of gauge theory, a future interpreter of both the positive and problematic contents of Schrödinger's *Space-Time Structure* would be able to say something akin to the following:

> All major advances of recent years (in General Relativity and Cosmology) have been made (partly) within this framework.[41]

In this section of this book, we now have two sets of observations: one about interpreting Weyl's results in history; and another about interpreting Schrödinger's work in history. Let's now bring both sets of observations together. For, either as explicitly present in O'Raifeartaigh's book, or by adverting to the historical context of Schrödinger's work, there are commonalities that foreshadow something of the full reach of the task of interpretation. That is, progress in reaching the meaning of an influential author's work can help us in our understanding of the entire geo-historical process that is all of physics.

This, however, raises a question: Does the task of interpretation include historical analysis? Is interpretation just an instance of historical analysis? For, an author's works emerge in a biography of years, and a biography of years is part of a geo-historical process of millennia. And so, interpretation needs to advert to historical reality. But, interpretation of *an* author's meaning is a focus on the works of *an* author within the geo-historical

---

[40] Merali [2013], p. 516.
[41] O'Raifeartaigh [1997], p. 242. Words in parentheses replace words from original text. See note 37.

community-body that is all of physics. And, the historical development of physics is, of course, not the work of any one person, but an integration of communications and collaborations in history that actually occurred. In other words, at least in provisional descriptive terms, historical development is partly determined by sequences of contributions of many authors, not only the works of authors, but works that in fact were taken up by the community. We have foreshadowings, then, that historical understanding, as such, is a distinct task from interpretation, although, certainly one that intrinsically depends on interpretation[42]. The task of historical understanding will be discussed in Chapter 3.

2.6   Future Heuristics for *Explanatory Interpretation*; and the Future Possibility of Interpreting Lonergan on Interpretation

Descriptive results in this book are superficial. Bernard Lonergan's reach was not. In Chapter 7 of the book *Method in Theology*[43], Lonergan briefly describes a few features of what he called *functional interpretation*. In particular, he draws attention to the need for "Understanding the Object," "Understanding the Words," "Understanding the Author" and "Understanding Oneself," respectively[44].

These might seem like straightforward recommendations regarding the task of interpretation. But, recall Section 2.4, regarding advanced content. And, note also that, in *Method*, Lonergan explicitly points out that "some such book as *Insight*" is a basic prerequisite for the book *Method*.

> I have presented this pattern of operations at length in the book, *Insight*. ... . But the matter is so crucial for the present enterprise that some summary must be included here. Please observe that I am offering only a summary, that the summary can do no

---

[42] Working out detailed heuristics of mutual dependencies among functional interpretation, functional history and other functional specialties, will be advanced work in later community progress. However, diagrams for preliminary heuristics are provided in Chapter 9.

[43] Lonergan [1975].

[44] Lonergan [1975], secs. 2, 3, 4, 5, respectively, pp. 156–162.

more than present a general idea, that the process of self-appropriation occurs only slowly, and, usually through a struggle with some such book as *Insight*. [45]

In *Method*, then, what does Lonergan mean by the task of interpretation? Obviously, we need to also take up his earlier work on interpretation, in Chapter 17 of *Insight*. I refer, in particular, to Section 17.3.8, Some Canons for a Methodical Hermeneutics[46]. That section, though, is extraordinary in its density, precision and reach. A high point in the section is a

canon of explanation [47],

regarding the possibility of *explanatory interpretation*:

The explanatory differentiation of the protean notion of being involves three elements. First, there is the genetic sequences in which insights gradually are accumulated by man. Secondly, there are the dialectical alternatives in which accumulated insights are formulated, with positions inviting further development and counterpositions shifting their ground to avoid the reversal they demand. Thirdly, with the advance of culture and of effective education, there arises the possibility of the differentiation and specialization of modes of expression; and since this development conditions not only the exact communication of insights but also the discoverer's grasp of his discovery, since such grasp and its exact communication intimately are connected with the advance of positions and the reversal of counterpositions, the three elements in the explanatory differentiation of the protean notion of being fuse into a single explanation.[48]

What is Lonergan getting at here? The paragraph is given in the 17th chapter of an advanced book of exercises, and so we can assume that its meaning is not to be reached without prior climbing. It is what McShane has identified as the crisis paragraph of the book *Insight*.[49] To add to the challenge of reading Lonergan's work on interpretation, Lonergan was, in

---

[45] Lonergan [1975], note 2, p. 7. See also, pp. 286–287; and Epilogue, Section E.2, Control of Meaning.

[46] Lonergan [1992], pp. 608–616.

[47] Lonergan [1992], p. 609.

[48] Lonergan [1992], 609–610.

[49] McShane [2013], pp. 118–119.

fact, a millennium class thinker who, among other things, was pushing the front lines of methodology[50].

Now, even without prior sections of the present chapter in mind, in the sciences it is obvious that in order to interpret an author's work, one at least needs to be up on the material. Einstein had his view of Weyl's 1918 paper.

> Einstein admired Weyl's (1918) theory "as a coup of genius of the first rate," but immediately realized that it was physically untenable. "Although your idea is so beautiful, that I have to declare frankly that, in my opinion, it is impossible that the theory corresponds to Nature.[51]

A few decades later, O'Raifeartaigh had his view of Weyl's work. Einstein's reading of Weyl's work did not rise to the standards of interpretation. But, he was able to understand and judge Weyl's papers. Certainly, this was partly because of Einstein's own command of physics of the time. And, how was it that O'Raifeartaigh was able to unpack Weyl's papers within an historical context? O'Raifeartaigh himself had contributed front-line work in field theory in the 20th century[52] and had "established himself as a researcher of the first rank."[53]

What, then, of the possibility of interpreting Lonergan's work, in particular, his second canon on interpretation in *Insight*[54]. Lonergan was a genius, certainly a scholar of the first rank with a nuanced grip on the development of physics and the sciences, mathematics and mathematical logic up to the 1950's, and philosophy and methodology of over two thousand years. Moreover, he had attained what so far in history is an extremely rare control of meaning.

Difficulties of interpreting Lonergan's work – for example, his work on interpretation – increase further still. For, while Lonergan climbed on from *Insight*, he did not offer many details in *Method* on implementation of functional specialization. Nor did he elaborate on how to incorporate his earlier work on interpretation (*Insight*, Chapter 17) with his later

---

[50] Lambert and McShane [2010], Part 3, Leading Ideas, pp. 165–254.

[51] O'Raifeartaigh and Straumann [2000], p. 1.

[52] Sen [2002].

[53] Sen [2002], par. 7.

[54] See note 48.

Chapter 7 in *Method*.[55] No doubt, though, he was aware of some of these difficulties, and the need for climbing to a control of meaning:

> Otherwise (the reader) will find not merely this chapter but the whole book about as illuminating as a blind man finds a lecture on color.[56]

Then, see also the listing he gives in *Method in Theology*, pp. 286–287[57], and where he briefly points to

> the contrast between differentiated consciousness that shifts with ease from one manner of operation in one world to another manner of operation in a different world and, on the other hand, undifferentiated consciousness which is at home in its local variety of common sense but finds any message from the worlds of theory, interiority, or transcendence both alien and incomprehensible.[58]

Is it not increasingly evident that interpretation of the work of Lonergan is a distant future possibility? Among other things, as is normal in all sciences, it will need collaboration among elite scholars who, in particular, have attained a mastery of material and a control of meaning at least comparable to what Lonergan achieved.

What, then, can we do now? We can take a page from how science makes progress. Even by today's standards, it is only a few who attempt interpretation, as such, of the works of influential authors. But, students and scholars still *learn* from great thinkers, and go on to make contributions to the field. There is a difference, then, between *learning* from a lead thinker and *interpretation* of the meaning of a lead thinker.

The academic community is not ready to interpret the work of

---

[55] Lonergan [1975], ch. 7, Interpretation. Some biographical information sheds additional light on the problem. Lonergan wrote *Method* in the late 1960's, at a time when he struggled with lung cancer. "(W)hen tired and sick, he faced in the late 1960's, the task of spelling out his findings, he shrunk his ambitions to addressing his Catholic community of thinkers, and he did so in light descriptive fashion. The address, unfortunately, fell on ears altogether familiar with descriptive talk, and in the main they missed the lion's lair of hints of a larger explanatory enterprise, a whole earth discipline, a program for at least seven millennia" (McShane [2012], p. 7).

[56] Lonergan [1975] p. 7.

[57] Lonergan [1975], pp. 286–287.

[58] Lonergan [1975], p. 287.

Lonergan. We can, though, learn from Lonergan. But, in that too there will be difficulties. In an effort to learn from Lonergan, readers might well attempt the self-ascent to which Lonergan invites us in *Insight*. In the book, he maps out a subtly ordered series of rapidly steepening exercises in self-attention in, among other areas, modern mathematics, mathematical logic, physics, and other sciences. The last paragraph of the first chapter of *Insight* gives a peek at just a few of the challenges involved:

> (O)ne may review the process from Euclidean to Riemannian geometry. Instead of asking why surds are surds, one can ask why transcendental numbers are transcendental. Similarly, one can ask whether the principle of inertia that Newton's laws are invariant under inertial transformations, what inspired Lorentz to suppose that the electromagnetic equations should be invariant under inertial transformations, whether an inverse insight accounts for the basic postulate of general relativity, whether the differences of particular places and particular times are the same aspect of the empirical residue as the differences of completely similar hydrogen atoms. For just as in any subject one comes to master the essentials by varying the incidentals, so one reaches familiarity with the notion of insight by modifying the illustrations and discovering for oneself and in one's own terms the point that another attempts to put in terms he happens to think will convey the idea to a probably nonexistent average reader.[59]

During my post-doctoral years (1992ff), among various projects in which I was involved, some colleagues and I had the pleasure of working through Peter Olver's Springer Graduate Series Text in Mathematics, *Applications of Lie Groups to Differential Equations*[60]. In a concise first chapter, the book briefly reviews the theory of manifolds, Lie groups, Lie derivatives and the de Rham complex. It then goes on to reformulate most of the classical and modern techniques and symmetry principles for obtaining exact solutions that were known up to the late 20th century. The book also includes generalizations of Noether's work as well as a chapter each on finite dimensional Hamiltonian Systems and Hamiltonian Methods, respectively. My colleagues and I very much enjoyed working through the book, cover to cover. We were able for the adventure in reading and learning (which necessarily included working through the exercises) thanks to previous teachers we had had and, in my case,

---

[59] Lonergan [1992], p. 56.
[60] Olver [1993].

background obtained in graduate school in mathematics as well as earlier studies in mathematical physics.

At this time in history, though, we do not yet have series of undergraduate and graduate texts that would prepare us – students and scholars – for reading *Insight*, a book that is a future graduate level (and beyond) text that might also be called *Self-Applications of Live Groupings of Operations*. It is, then, quite the (worthwhile) fantasy to think of future explanatory interpretations of Lonergan's book *Insight*, not to mention his later writings on functional interpretation within functional specialization. In the meantime, we can make the effort to learn from the master. And, that learning will be helped greatly through efforts in implementing his later discovery that there are eight main tasks in the Academy.

2.7 References

Baez, J. (2012) *Open Questions in Physics* (http://math.ucr.edu/home/baez/physics/).

Einstein, A. (1987) *The Collected Papers of Albert Einstein*, eds. Stachel, J., Cassidy, D. C. and Schulmann R. (Princeton University, Princeton, NJ).

Hawking, S. and Ellis, G. F. R. (1973) *The large scale structure of space-time* (Cambridge University Press, Cambridge).

Jackson, J.D. and Okun, L. B. (2001) Historical roots of gauge invariance, *Rev. Mod. Phys.* 73, pp. 663–680.

Kykhen, J., and Spiropulu, M. (2014) Sypersymmetry and the Crisis in Physics, *Sci. Am.*, May, pp. 34–39.

Lambert, P. and McShane, P. (2010) *Bernard Lonergan. His Life and Leading Ideas* (Axial Publishing, Vancouver).

Lewis, J. T., McGlinn, W. and Sen, S. (2001) Obituary of Lochlainn O'Raifeartaigh, *Phys. Today*, Nov., p. 79.

Lonergan, B. (1975) *Method in Theology* (Darton, Longman & Todd, London).

Lonergan, B. (1992) *Insight: A Study of Human Understanding*, vol. 3 in *Collected Works of Bernard Lonergan*, Crowe, F. E. and Doran, R. M., eds. (University of Toronto Press, Toronto).

McShane, P. (2001) Elevating *Insight*: Space-Time as Paradigm Problem, *Method: J. Lonergan Studies*, 19, pp. 203–229. Or, for free online access, see http://www.philipmcshane.org/published-articles/.

McShane, P. (2012) *Method in Theology 101 A.D. 9011: The Road to Religious Reality* (Axial Publishing, Vancouver).

McShane, P. (2013) *Futurology Express* (Axial Publishing, Vancouver).

Merali, Z. (2013) The Origins of Space and Time, *Nature*, vol. 500, August, pp. 516–519.

Olver, P. (1993) *Applications of Lie Groups to Differential Equations, Graduate Texts in Mathematics*, vol. 107 (Springer-Verlag, New York).

O'Raifeartaigh, L. (1997) *The Dawning of Gauge Theory* (Princeton University Press, Princeton, NJ).

O'Raifeartaigh, L. and Straumann, N. (2000) Gauge theory: Historical origins and some modern developments, *Rev. Mod. Phys.*, vol. 72, No. 1, January, pp. 1–23.

Schlosshauer, M., Kofler J. and Zeilinger, A. (2013) A snapshot of foundational attitudes toward quantum mechanics, *Stud. Hist. Philos. of Mod. Phys.*, vol. 44, issue 3, pp. 222–230.

Schrödinger, E. (1950) *Space-Time Structure* (Cambridge University Press, Cambridge).

Sen, S. (2001) Lochlainn O'Raifeartaigh 1933–2000, http://www.stp.dias.ie/Lochlainn/lochlainn.html.

Smolin, L. (2006) *The Trouble with Physics, The Rise of String Theory, the Fall of a Science, and What Comes Next* (Houghton Mifflin, Boston).

Smolin, L. (2013) *Time Reborn: From the Crisis in Physics to the Future of the Universe* (Houghton Mifflin Harcourt, Boston).

Chapter 3

Functional History

3.1 Introduction

In this chapter, I look to a few examples of historical work in physics, most of which (but not all) are about field theories of the 20th and 21st centuries. Certainly, this chapter is not intended to provide a representative sample of historical works in physics. That task needs global collaboration. This chapter does, though, provide a sampling. And, taking help from Lonergan's pointers on historical method, it is possible to discern some elements of historical work that will eventually luminously be part of a progress-oriented historical analysis that in this book is called *functional history*. Taking O'Raifeartaigh's words, the sampling "is not comprehensive. ... The choice ... is to give the flavor"[1] of the problem.

The structure of the chapter is as follows: Section 3.2.1 looks again to Lochlainn O'Raifeartaigh's book,[2] this time concentrating on aspects of his book that explicitly contribute to historical understanding. Section 3.2.2 includes a few comments on a paper by Jackson and Okun[3] regarding the historical origins of gauge invariance. Section 3.2.3 is on Smolin's[4] A Brief History of String Theory, Part II of his book *The Trouble with Physics*[5]. Section 3.2.4 is on a book by Helge Kragh[6], after which, in Section 3.2.5, I comment only indirectly on *The Oxford Handbook of the*

---

[1] O'Raifeartaigh [1997], p. viii.
[2] O'Raifeartaigh [1997].
[3] Jackson and Okun [2001].
[4] Smolin [2007].
[5] Smolin [2006].
[6] Kragh [1999].

History of Physics.[7] In footnotes, other works are added to the context. Section 3.3 invites attention to a few key issues that eventually will need to be resolved. Section 3.4 is to help us get some sense of the shape of a future functional history. This is done by combining our preliminary observations with some of Lonergan's pointers on historical method.

3.2 Preliminary Descriptions

3.2.1 *The Dawning of Gauge Theory*[8]

In Chapter 2, *The Dawning of Gauge Theory* was used to illustrate elements of interpretation. But, the book includes much more, for in it the

early history of gauge theories has been beautifully recounted[9].

More specifically, in addition to interpretation, O'Raifeartaigh's book also includes elements of historical analysis that are progress-oriented. So, in this subsection, I touch briefly on three main questions[10]: (1) What was the historical component, as such, in what O'Raifeartaigh was doing? (2) How did he do it?; and (3) In what ways is the historical work in the book intended to contribute to progress?

In order to say something about (1), I begin with samples of what O'Raifeartaigh himself says regarding the purpose of his book, but I also include some brief descriptions of some parts of the text.

In the Preface, O'Raifeartaigh writes:

- The purpose of this book is to give a short sketch of the first two lesser-known stages (in the development of gauge theory).[11]
- Indeed one motivation for writing this book was to make this early work on

---

[7] Buchwald and Fox [2013].
[8] O'Raifeartaigh [1997].
[9] Taylor [2001], p. xi.
[10] Certainly, these questions are not independent of each other. What would be needed is interpretation of O'Raifeartaigh's work: see ch. 2.
[11] O'Raifeartaigh [1997], p. viii.

gauge theory more accessible by assembling some of the relevant articles and translating them into English. [12]

• It is hoped that the book will be of value not only as description of how modern gauge theory developed in its early days but also in showing how original ideas develop and can come to fruition in spite of initial difficulties and frustrations. [13]

From the Introduction, there are additional comments:

• The purpose of the book is to explain how this (so-called *gauge*) structure gradually emerged. [14]

• Each of the (two) phases in the development of gauge theory has an interesting history, and the present book is an attempt to give a brief description of the first phase (the first two stages stages), which is perhaps less well-known. [15]

The book has two parts. Part I covers the development of gauge theory from Einstein's theory of gravitation to Weyl's 1929 paper (what O'Raifeartaigh calls the first stage). Chapter 1 of the book also gives general historical background. As a supplementary result, O'Raifeartaigh includes one of Feynman's unpublished calculations[16], which was an

---

[12] O'Raifeartaigh [1997], p. viii.

[13] O'Raifeartaigh [1997], p. ix.

[14] O'Raifeartaigh [1997], p. 3.

[15] O'Raifeartaigh [1997], p. 4. "The second phase, which lasted roughly from the mid-sixties to the mid-eighties, was concerned with the application of the existing theory to nuclear forces, whose structure had by that time become much clearer" (O'Raifeartaigh [1997], p. 4).

[16] "(There is) the question as to whether or not electromagnetism, like gravitation, could be derived from some more universal principle. No such principle is known at present but it might be interesting to mention some earlier work of Feynman that has recently been published and throws some light on the question. Feynman was ... looking for some new type of fundamental force and for this purpose decided to investigate what would happen if the canonical commutation relations (or Poisson brackets) for the momentum variables p were relaxed. To his surprise and disappointment, he found that the force obtained in this way is just the conventional electromagnetic force and did not publish his results. But, for historical reasons, and because the result could be interpreted positively as a derivation of electrodynamics, the Feynman argument was recently published by Dyson" (O'Raifeartaigh [1997], pp. 17–18). Dyson's article is, Dyson [1990]. A small literature has emerged from Dyson's publication of Feynman's result. See, e.g., *Am. J. Physics*, 1990 and following. Articles have included technical comments on mathematical derivations in various contexts, including both pre-relativistic and relativistic physics. There is also,

attempt to derive electromagnetism from a universal principle[17]. In the other four chapters of Part I, key articles from the tradition are reproduced. Each chapter includes commentary regarding corresponding articles.

Part II is on what O'Raifeartaigh calls the second stage in the development of gauge theory. It treats generalizations of Weyl's gauge formulation of electromagnetism to non-abelian gauge theory. It contains articles of Klein, Pauli, Yang and Mills, Shaw and Utiyama. Each chapter in Part II again is accompanied by commentary on the background, history and various motivations that led up to each of the articles.

Where the main body of *The Dawning of Gauge Theory* focuses on the first two stages (both together what O'Raifeartaigh calls the *first phase*) in the development of gauge theory, the Conclusion[18] of the book is a (highly abbreviated) summary of the third stage (what O'Raifeartaigh calls the *second phase*).

> From the ... brief summary it will be clear that the developments in gauge theory subsequent to 1954 were just as complicated and fascinating as those prior to that date. But, their proper recounting is beyond the scope of this book.[19]

Nevertheless, that is not O'Raifeartaigh's last word on the later development.

> (T)he geometrization of gravity raised the question as to whether these other fundamental (electromagnetic and nuclear) forces were 'true' forces operating in the curved space of gravitational theory or whether they also were part of the geometry.

Bérard [2007]. More recently, there is an interpretive piece of Feynman's derivation, given from a philosophical viewpoint: Pombo [2009].

[17] Was there a missed opportunity? *Contra-factual history* is an emerging focus. For the historian, events in history are provisionally identified relative to a present view of progress and decline. See Section 3.4. By the same token, missed opportunities and 'roads not taken' also are known relative to an *acquis*. For some discussion of these issues, see, e.g., Pessoa [2001] and references therein. However, contra-factual history is in its infancy. In particular, the literature is not yet in control of differences between (past-oriented) contra-factual history and the (future-oriented) task of developing views of real possibilities. The community task of developing views of real possibilities will be discussed in ch. 7, Functional Systematics.

[18] O'Raifeartaigh [1997], "Conclusion," pp. 240–242.

[19] O'Raifeartaigh [1997], p. 241.

This question has still not been fully answered. But what has become clear is that these forces and gravitation have a common geometrical (*gauge*) structure.[20]

In the last two paragraphs of the book, he writes:

The elevation of the gauge fields to the level of the gravitational field is a substantial achievement, but it is by no means the end of the story. Indeed, there are two major limitations on the power of gauge theory. First, in its present form at least, it does not unify the fundamental interactions in an intrinsic way, in the sense that the coupling constants for the different interactions remain theoretically undetermined. … Furthermore, gravitation is not unified with the other interactions, … . Second, gauge theory provides no answers for the questions that arise concerning the matter fields, such as the distinction between baryons and leptons, the existence of three generations, of quark-lepton pairs, the origin and structure of the symmetry-breaking scalar sector, and of the quark mixing matrix.[21]

O'Raifeartaigh also includes a final comment that looks toward future progress:

What gauge theory does is to determine the interaction of the radiation fields themselves and with matter, and to provide a universal and reliable frame-work for further investigation and development.[22]

Such, then, is a brief sketch of some of the main contents of O'Raifeartaigh's book. Recall, now, question (1): What was the historical component, as such, in what O'Raifeartaigh was doing? Evidently, he is not simply looking back to earlier times and results. Nor is he merely organizing names, dates and technical results[23]. Nor does the book consist

---

[20] O'Raifeartaigh [1997], p. 3.

[21] O'Raifeartaigh [1997], p. 242.

[22] O'Raifeartaigh [1997], p. 242.

[23] A helpful resource written along these lines is Goenner [2004]. The article is on the period 1914–1933, "when Einstein was living and working in Berlin" (Goenner [2004], p. 1). The article is book-length (127 pages), has 24 pages of references with 434 works cited, and provides extensive detail on authorship and chronological orderings. "It includes brief technical descriptions of the theories suggested, short biographical notes concerning the scientists involved, and an extensive bibliography" (Goenner [2004], p. 1). As the author intended, there is no analysis of developments. The book points to an extensive range of potentially significant data: "Even a superficial survey such as the one made here shows clearly the dense net of mathematicians and theoretical physicists involved in the building of unified field theory and of the geometric structures underlying it" (Goenner [2004],

merely of a collection of historically significant articles (in translation, when needed)[24]. As O'Raifeartaigh says for himself, he wants to *explain*

the emergence of gauge theory ... a gradual process, a slow evolution rather than a revolution.[25]

And, since

some of (the) papers are difficult to peruse from a modern point of view[26]

he also

attempted to summarize the salient points.[27]

But, of course, he does not merely summarize. O'Raifeartaigh helps the modern reader get inside the early development of gauge theory. He describes Weyl's first unsuccessful attempt to unify gravity and electromagnetism. He includes some of Einstein's reflections on Weyl's 1918 paper,

a coup of genius[28],

---

p. 125). Goenner's article also helps draw attention to further dimensions of historical analysis: "While my focus lies on the conceptual development of the field, by also paying attention to the interaction of various schools of mathematicians with the research done by physicists, some prosopocraphical remarks are included" (Goenner [2004], p. 1). Note: *prosopography*: a study that identifies and relates a group of persons or characters within a particular historical or literary context, *Merriam-Webster English Dictionary,* 2016.

[24] One such collection is Taylor [2001]: "My intention in this volume is to bring together some of the key papers in the development of gauge theories in physics, mainly from the 1920's to the 1980's. Inevitably, some equally important papers will have been omitted. The emphasis is on principles rather than applications. The volume is not intended to be a serious work in the history of science; still less is it a textbook of physics" (Taylor [2001], p. ix). However, such a collection might well be of interest to future functional researchers. See ch. 1.

[25] O'Raifeartaigh [1997], p. 3.

[26] O'Raifeartaigh [1997], pp. 9–10.

[27] O'Raifeartaigh [1997], pp. 9–10.

[28] For bibliographical data on primary sources, see O'Raifeartaigh and Straumann [2000], p. 1. See also, Straumann [1996].

but, as Einstein noted, it lacked correspondence with Nature.[29] He then goes on to identify subtleties in advances made by authors, details of transitions as well as ongoing confusions.

In the Conclusion of his book, O'Raifeartaigh further contextualizes his results within the modern context (1997), and brings new focus to open questions. His results also led him to comment more generally on development:

> It is hoped that the book will be of value not only as description of how modern gauge theory developed in its early days but also in showing how original ideas develop and can come to fruition in spite of initial difficulties and frustrations.[30]

For each of the articles O'Raifeartaigh discusses in his book, there is, of course, an interpretation problem[31]. But, O'Raifeartaigh does more than provide a collection of unrelated interpretations. What else is he doing? He is describing and relating transitions, connections and differences between understandings, ongoing confusions and limitations in the works of Weyl; Kaluza; and so on. In other words, further to the task of interpreting the works of individual authors, O'Raifeartaigh is sorting out what in fact was moving forward, for better or for worse, advances as well as confusions that in fact were taken up by the community and that, in particular, eventually led to the emergence of gauge theory as we know it.

O'Raifeartaigh made progress in identifying complexly structured sequences. Which sequences? There are sequences of contexts, theories;

---

[29] "If (the implications of Weyl's hypotheses) were so in Nature, chemical elements with spectral-lines of definite frequency could not exist and the relative frequency of the two neighboring atoms of the same kind would be different in general. As this is not the case it seems to me that one cannot accept the basic hypothesis of this theory, whose depth and boldness every reader must nevertheless admire" (Albert Einstein, "Postscript, A Remark by Mr. A. Einstein," addendum to Weyl's 1918 paper, reproduced in full, along with the published exchange with Einstein, in O'Raifeartaigh [1997], pp. 24–37). For primary sources, see Stachel, Cassidy and Schulmann [1987]. For open access as well as English translations of Einstein's papers, see http://einsteinpapers.press.princeton.edu/.

[30] See note 13.

[31] Regarding the interpretation problem, see ch. 2, above.

questions, motivations; insights, understandings and confusions in the works that contributed to what in fact transpired in the community.[32]

This brings me now to question (2): How did O'Raifeartaigh do all of this? He speaks about what Weyl and others understood. He speaks about subtle differences and shifts in understanding, with tremendous technical control over the material. How, though, can he speak with understanding, about differences and shifts in understanding between Weyl and others?

Just as in interpretation, there is what in some sense may seem obvious but is not often adverted to, namely: the understandings O'Raifeartaigh speaks about are his own. In order to speak with understanding about each of the key papers, let alone sequences of transitions between papers, O'Raifeartaigh himself understood the relevant physics. But, it was not enough for him to simply understand gauge theory and other relevant physics. In addition to his understanding in gauge theory, he inquired into the evolution[33] of gauge theory. And, in that further inquiry, he needed to identify sequences of understandings in his own understanding; issues that might, in his estimation, have the potential for leading to confusion; and issues that, to his mind, might lead to future developments. The entire book is, after all, Lochlainn O'Raifeartaigh writing about his understanding of the development of gauge theory, and explaining what to his mind was going forward in the physics community, for better or for worse.

Let's now look to question (3). In what ways is the historical work in O'Raifeartaigh's book intended to contribute to progress? With a focus on gauge theory, O'Raifeartaigh sheds new light on the growth of physics so far (up to *circa* 1997). His work is an expression of his own growth in the gauge-theory-focused question, 'How have we been growing?'

But, his aim is to do more than merely help us better understand prior growth, although that in itself is a kind of growth. He also identifies two

---

[32] O'Raifeartaigh [1997] does not make this a point of discussion. However, this is evident in the book; and was diagrammed in an article written soon after, with co-author Norbert Straumann. See, "Figure 1. Key papers in the development of gauge theories" in O'Raifeartaigh and Straumann [2000], p. 2. This same diagram also appeared previously in Straumann [1996]. The relevance of sequences in understanding historical developments also is alluded to by Goenner: There is "the dense net of mathematicians and theoretical physicists involved in the building of unified field theory and of the geometric structures underlying it" (Goenner [2004], p. 125; see also note 23).

[33] See note 25.

major limitations of gauge theory.[34] He goes on to relate his results on the development of gauge theory to (what for him was) the contemporary (1997) context in field theory. O'Raifeartaigh points out that gauge theory provides a frame-work for further investigation.[35] He indicates that his work might help future progress in theoretical physics, a prior task which, in fact, was a main focus of his career.

> The elevation of gauge fields to the level of the gravitational field is a substantial achievement, but is by no means the end of the story.[36]

However, O'Raifeartaigh also hopes that his book will shed light on how ideas develop and come to fruition.[37] In that sense, he points beyond the context of historical understanding of any particular gauge theories. He indicates that his work might contribute to an improved understanding of scientific development itself[38]. By the end of the book, O'Raifeartaigh is asking questions about progress. His questions are not about particular transitions and shifts in the evolution of gauge theory. Instead, they call for collaboration in further tasks that will be discussed in subsequent chapters of this book.

3.2.2 *Historical Roots of Gauge Invariance*[39]

A few years after O'Raifeartaigh's book, an article was published called "Historical roots of gauge invariance."[40]

> A number of authors have discussed the ideas and history of quantum gauge theories, beginning with the 1920's, but the roots of gauge invariance go back to the year 1820 when electromagnetism was discovered and the first electrodynamic theory was proposed. We describe the 19th century developments that led to the discovery that different forms of the vector potential ... are physically equivalent.[41]

---

[34] See note 21.

[35] See note 22.

[36] O'Raifeartaigh [1997], p. 242.

[37] See note 30

[38] See note 13.

[39] Jackson and Okun [2001].

[40] Jackson and Okun [2001].

[41] Jackson and Okun [2001], p. 663.

As described in the abstract of their paper, a main focus of the article by Jackson and Okun is on understandings of *potential functions* in electrodynamics, prior to Weyl's 1929 article[42]. The paper does not include analysis, but does provide a detailed record of names, dates, formulas and some of the experiments that contributed to the historical development. The paper cross-references these results with other historical works. The last section of their paper, Section IV, however, changes focus. It is not on the origins of gauge theory, but speaks to the Physical Meaning of Gauge Invariance. The section also expresses something of the authors' basic views implicit in their work as historians.

3.2.3 *The Trouble with Physics*[43]

In *The Trouble with Physics*, Lee Smolin takes himself and the physics to community to task for having allowed string theory to become what at present is

the dominant paradigm[44]

in departments of theoretical physics. It is not that Smolin argues against string theory having a place in contemporary theoretical physics:

(S)tring theory is certainly among the directions that deserve more investigation.[45]

In fact, earlier in his career, Smolin himself published work that contributed to string theory. But, more recently, he writes that

string theory has not been successful enough on any level to justify putting all our eggs in its basket.[46]

---

[42] "Historically, of course, Weyl's 1929 papers were a watershed" (Jackson and Okun [2001], p. 676).
[43] Smolin [2006].
[44] Smolin [2006], p. 199.
[45] Smolin [2006], p. 199.
[46] Smolin [2006], p. 199. August 2016 experimental results from the LHC remain consistent with the claim. See, for example, Gibney [2016].

Reaching for a perspective on string theory is, however, only one part of *The Trouble with Physics*. Chapter 1 of the book begins with a discussion of

the five great problems in theoretical physics[47],

while the last chapter, Chapter 20, includes, among other things, practical recommendations for both the scientific community and the general public. There are four parts to the book: Part I, The Unfinished Revolution (chs. 1–6), leads up to the 1980's, at which time "there had been no progress on making a theory of quantum gravity."[48] Part II is A Brief History of String Theory (chs. 7–12). Part III, Beyond String Theory (chs. 13–15) looks to new data in physics, as well as recent areas of development besides string theory. Part IV, Learning from Experience (chs. 16–20)

return(s) to the questions ... raised in the Introduction. Why, despite so much effort by thousands of the most talented and well-trained scientists, has fundamental physics made so little definitive progress in the last twenty-five years? And given that there are promising new directions, what can we do to ensure that the rate of progress is restored to what it was for two centuries before 1980?[49]

The Trouble with Physics is, in a sense, deceptively readable. For it is, in fact, a complex work about developments in 20th century physics and, no doubt, will need to be recycled by future scholars. It is, after all, the work of a contemporary lead theoretical physicist providing something of a

360-degree look[50]

at the problem of progress in contemporary physics.

---

[47] Smolin [2006], ch. 1, pp. 3–17.

[48] Smolin [2006], p. 98.

[49] Smolin [2006], p. 261.

[50] Smolin [2006], p. 335. I am taking the phrase out of context. The quotation is from where Smolin recalls performance assessment methods in businesses outside of the academy. But, the phrase "360-degree look" gives some sense of the past-and-future scope of Smolin's book. See, for example, What We Can Do for Science, Smolin [2006], ch. 20, pp. 349-355.

A further difficulty is that the book is multi-tasking. It looks both to the past and to the future. It appeals to a variety of sources, and types of sources, and includes various aims. Smolin's book, as it happens, provides data on several functionally distinct tasks. In Chapter 4, I will return to Smolin's book, because parts of his book reveal a focus within a fourth task. For the present Chapter 3 on functional history, I draw from Part II[51], a history of string theory up to *circa* 2005.

I select just a few samples from Part II. These are not intended in any way to summarize Part II. They are to show a few of the different foci. Each sample is followed by my *Comments*. At the end of the collection of samples and *Comments*, I speak briefly to Part II as a whole.

Part II, A Brief History of String Theory, begins with Chapter 7, Preparing for a Revolution. In the first pages of the chapter, Smolin recalls some details of the (well-known) beginnings of string theory:

> In 1968, a young Italian physicist named Gabriele Veneziano saw an interesting pattern in the data (obtained from collisions of strongly interacting particles). He described the pattern by writing down a formula that described the probabilities for two particles to scatter from each other at different angles. Veneziano's formula fit some of the data remarkably. It caught the attention of some of his colleagues, in Europe and in the United States, who puzzled over it. By 1970 a few were able to interpret it in terms of a physical picture. … *string theory* that was born.[52]

Comments: In pages 103-104[53] of his book, Smolin gives a partial description of Veneziano's formula. It seems that Nambu, Nielsen and Susskind were not in close contact. Their articles reveal three relatively independent developments emerging from Veneziano's formula: Veneziano to Nambu; Veneziano to Nielsen; and Veneziano to Susskind. Dates and locations are given. Note that, in addition to speaking of technical developments, Smolin also recalls some of the difficulties that Susskind encountered, when trying to communicate the new ideas to the

---

[51] Smolin [2006], pp. 99–199.
[52] Smolin [2006], pp. 103–104.
[53] Similar page references to Smolin's book are in *Comments*, below.

community. The biographical detail also reveals something of prevailing views in the physics community at that time.

In 1972, Neveu and another French physicist, Joël Scherk, found that the superstring had states of vibrations corresponding to gauge bosons, including the photon. This was a step in the right direction. But an even bigger step was taken two years later, by Scherk and Schwarz. They found that some of the massless particles predicted by the theory could actually be gravitons. (The same idea occurred independently to a young Japanese physicist, Tomiaki Yoneya.)[54]

Comments: On p. 106, Smolin gives a few details of further terms in sequences contributing to development of the new theory.

The First superstring revolution took place in the fall of 1984.

...

String theory promised what no other theory had before – a quantum theory of gravity that is also a genuine unification of forces and matter. It appeared to offer, in one bold beautiful stroke, a solution to at least three of the five great problems of theoretical physics.[55]

Comments: In pp. 114–115, Smolin describes technical advances obtained by Green and Schwartz, and relates that work to the five great problems he noted in Chapter 1 of his book. He also describes something of the hopes of the string theory community at that time.

Very quickly there developed an almost cultlike atmosphere. You were either a string theorist or you were not.[56]

...

Mathematics now sufficed to explore the laws of nature. We had entered the period of postmodern physics.[57]

---

[54] Smolin [2006], p. 106.
[55] Smolin [2006], pp. 114–115.
[56] Smolin [2006], p. 116.
[57] Smolin [2006], pp. 116–117.

Comments: In pp. 116–117, deviations from standard practice are indicated, in methods and in ethos of the new theory.

All the string theories predicted extra particles – particles not seen in nature.[58]

Four leading string theorists (Philip Candelas, Gary Horowitz, Andrew Strominger, and Edward Witten)

were able to show that the conditions needed for string theory to reproduce a version of the supersymmetric standard model were the same as the conditions that defined a Calabi-Yau space.[59]

...

Among the detractors was Richard Feynman, who explained his reluctance to go along with the excitement as follows:

I don't like that they're not calculating anything. I don't like that they don't check their ideas. I don't like that for anything that disagrees with experiment, they cook up an explanation – a fix up to say "Well, it still might be true." ... So the fact that it might disagree with experience is very tenuous, it doesn't produce anything; it has to be excused most of the time. It doesn't look right.

These sentiments were shared by many of the older generation of particle physicists, who knew that the success of particle theory had always required a continual interaction with experimental physics.[60]

Comments: In pp. 119–125, Smolin describes an increasingly complex situation in the theoretical physics community. He notes the emergence of "great opportunities, and great problems"[61] in string theory, and recalls the breakthroughs made by Candelas, Horowitz, Strominger, and Witten. But, Smolin also calls attention to an increasing strain between string theory groups that did not call for experimental evidence, and "older generation" physicists who insisted that ideas eventually agree with experiments. In pp. 127–128, Smolin also draws attention to discrepancies between technical progress and available data. He describes an uneasy division

---

[58] Smolin [2006], p. 121.
[59] Smolin [2006], p. 123.
[60] Smolin [2006], p, 125.
[61] Smolin [2006], p. 119.

emerging in the community – between "believers" in string theory and "skeptics" (those who insisted that a physical theory be in some way experimentally verifiable).

The second superstring revolution, which burst on the scene in 1995, gave us just that. The birth of the revolution is often taken to be a talk that Edward Witten gave that March at a string theory conference in Los Angeles, where he proposed a unifying idea.[62]

...

Later that year, Witten gave the so-far-undefined theory a name. ... he called it simply *M-theory*.[63]

...

There have been some fascinating hints, but we still do not know what M-theory is, or whether there is any theory deserving of the name.[64]

...

Unfortunately, M-theory remains a tantalizing conjecture. It's tempting to believe it. At the same time, in the absence of a real formulation, it is not really a theory – it is a conjecture about a theory we would love to believe in.[65]

Comments: In pp. 129–147, Smolin recalls Witten's M-theory. He notes a problem, namely, that M-theory is not a theory yet, but a conjecture regarding a possible not-yet-obtained theory.

In the two string revolutions, observation played almost no role.

...

---

[62] Smolin [2006], p. 129.
[63] Smolin [2006], p. 136.
[64] Smolin [2006], p. 146.
[65] Smolin [2006], p. 147.

The long-held hopes for a unique theory have receded, and many of them now believe that string theory should be understood as a vast landscape of possible theories, each of which governs a different region of a multiple universe.[66]

Comments: In pp. 149–150, Smolin again draws attention to the historical result that observation has played almost no role in string theory. In particular, string theorists did not attempt to resolve the problem of non-uniqueness but, instead, appealed to belief in string theory. Within the string theory community ethos, this led to conjectures about a multi-verse.

The answer depends on what value of the cosmological constant we want to come out. If we want to get a negative or zero cosmological constant, there are an infinite number of distinct string theories. If we want the theory to give a positive value for the cosmological constant, so as to agree with observation, there are a finite number; at present there is evidence for 10^{500} or so such theories.[67]

Comments: In pp. 157–158, Smolin reviews some details in string theorists' understanding of the cosmological constant. Additional technical problems with string theory are noted. In particular, 10^{500} or so geometries could satisfy the basic premises, with no apparent way to distinguish them experimentally.

(String theories require) special choices of background geometries, and this contradicts Einstein's principal discovery, as set out in his general theory of relativity, that the geometry of spacetime is dynamical and that physics must be expressed in a background-independent manner.

...

The second generic prediction of string theory is that world is supersymmetric.

...

Unfortunately, because supersymmetric theories have so many free parameters, there is no specific prediction of what the mass of the WIMPs (weakly interacting massive particles) should be or exactly how strongly they should interact.

---

[66] Smolin [2006], pp. 149–150.
[67] Smolin [2006], pp. 157–158.

...

Experimentalists have looked for WIMPs by using detectors similar to those used to detect neutrinos coming from the sun and distant supernovas. Extensive searches have been carried out, but so far no WIMPs have been found.

...

We now come to the third generic prediction of string theory: that all the fundamental forces become unified on some scale.

...

Experiments have looked for proton decay and failed to find it. These results (or lack of them) have killed certain grand unified theories but not the general idea. However, the failure to find proton decay remains a constraint on possible theories, including supersymmetric string theories.[68]

Comments: In pp. 174-176, Smolin draws attention to three main sets of difficulties with string theory: inconsistencies with Einstein's work; the fact that no weakly interacting particles predicted by string theories have been detected; and, nor has there been any evidence for proton decay.

Note that the samples above from Part II are from chapters 7–11 only. No samples are taken from Chapter 12, the last chapter of A Brief History of String Theory. This is because Chapter 12, What String Theory Explains, includes questions with a different focus. For example:

What are we to make of the strange story of string theory so far? ...Mindful that the future is always open, I would like to close this section[69] by offering an *assessment*[70] of string theory as a proposal for a scientific theory.[71]

I defer discussion of Smolin's Chapter 12, then, to Chapter 4 of the present book. What, though, of chapters 7–11 of *The Trouble with Physics*? Similar to Section 3.2.1 on O'Raifeartaigh's work, I organize discussion here around – in this case, two – main questions: (1) What is Smolin doing in the five chapters 7–11 of A Brief History of String

---

[68] Smolin [2006], pp. 174–176.
[69] "section": that is, Smolin [2006], Part II, A Brief History of String Theory, chs. 7–12.
[70] Italics mine.
[71] Smolin [2006], p. 177.

Theory? (2) In what way are these five chapters intended to contribute to progress in the physics community?

Regarding (1): Part II of *The Trouble with Physics* tracks origins and theoretical developments leading up to contemporary string theory. Chapters 7–11 tell something of motivations, mathematical discoveries, and technical difficulties overcome. They also draw attention to what, in Smolin's view, are various deviations from what, historically, has been standard practice in physics. Not being exhaustive here, deviations noted include: an emphasis on mathematical development without requiring experimental evidence; admitting fundamental ambiguities within string theory itself (e.g., non-uniqueness), and that available computations actually are through approximation theories to unknown theories; that the few predictions that have been available from approximations are not compatible with either known experimental results or general relativity.

So, the focus of Part II varies considerably. There are observations regarding mathematical development; experimental results and lack thereof; scientific method; community ethos; and community tensions. The five chapters 7–11 historically contextualize layerings of details in varieties of sequences of understandings and implementations in the physics community. In other words, Part II of Smolin's book is an extremely complex weave that, in a single telling, points to many types of question and result, connectivities in sequences of developments, as well as what, in Smolin's view, are both positive and negative (or at least problematic) developments in the history of string theory.

Regarding (2): Even though 'brief,' chapters 7–11 of *The Trouble with Physics* expresses a new (2005) descriptive understanding of string theory and its development up to *circa* 2005. It does so partly by looking to the works of individuals, transitions made between works of individuals, and also to trends in the whole community. No detail seems to be too small and no trend too large, if potentially helpful in understanding the development. As the book makes explicit, the development has drawn on and affected all of physics. "A Brief History of String Theory" is, then, a contribution toward new understanding not only of the development of string theory, but of what has been going forward (or not), of what in fact happened (or not), for better or for worse, in string theory, within and relative to the physics community.

Part II, however, is at some remove from sources. The historical account points to various technical details. It does not, however, attempt to identify differences in types of result, the nature and details of the various transitions, nor does it go into details regarding methodological issues.

The book will be accessible to physicists. And, because the many observations made by Smolin regard advances in complex combinations of description, theory and (implicitly) philosophies, the work might well invite recycling. It could serve as a potential resource for future (functional) research.

3.2.4 *Quantum Generations*[72]

The book *Quantum Generations* is a history about physics in the 20th century. From the Preface of the book, Kragh writes:

> What follows is a ... fairly brief and much condensed and selective account of what I believe have been the most significant developments in a century of physical thought and experiment.
>
> ...
>
> The book's goal is to give an account of the development of physics during a one-hundred year period that is digestible, informative, and fairly representative. [73]

In somewhat more detail:

> Taken together, the twenty-nine chapters cover a broad spectrum of physics, not only with respect to topics and disciplines, but also with respect to the dimensions of physics. ... The present work is not devoted solely to the scientific or intellectual aspects of physics, but neither does it concentrate on the social and institutional history. It tries to integrate the various approaches, or at least, to include them in a reasonably balanced way. I have also paid more attention to applied or engineering physics than is usually done. To ignore the physics-technology interface, and concentrate on so-called fundamental physics alone, would surely give a distorted picture of how physics has developed in this century. Not only are many of the world's physicists occupied with applied aspects of their science, and have been so

---

[72] Kragh [1999].
[73] Kragh [1999], p. xi.

during most of the century, but it is also largely through the technological applications that physics has become a major force of societal change.[74]

Later in the book, Kragh writes:

> In terms of manpower, organization, money, instruments, and political (and military) value, physics experienced a marked shift in the year following 1945. The sociopolitical changes made physics in 1960 a very different science than it had been a century earlier.[75]

But, what kind of history does Kragh provide?

> I have described. rather than analyzed, important parts of the development of physics between 1895 and 1995.[76]

More precisely, the content of the book is largely determined by questions given at the beginning of Chapter 2, The World of Physics[77]:

> Who were the physicists around 1900? How many were there and how were they distributed among nations and institutions? What kind of physics did they work with? How were they funded?[78]

These questions carry through the rest of the book. Statistical results are given from 1895–1995. These include: numbers of physicists in areas of research; numbers of Ph.D. students; distributions of topics in publications; page counts devoted to different areas; physics institutes and faculty; physics in industrial laboratories; topics and contributions by country; distributions of CERN personnel; funding sources and amounts, including military funding; Nobel prizes in physics, 1901–1998. Much of this is recorded in statistical tables and figures.

The statistical data, though, are only part of the book. Kragh notes major authors, their results, points to transitions in theoretical development, as well as controversies (for example, those around the

---

[74] Kragh [1999], p. xii.
[75] Kragh [1999], p. xiv.
[76] Kragh [1999], p. xiv.
[77] Kragh [1999], The World of Physics, ch. 2, pp. 13–26.
[78] Kragh [1999], p. 13.

emergence of string theory[79]). He also mentions cultural, social and economic factors in the history. In the last chapter of the book, Kragh provides a number of concluding remarks, two of which are:

> Physics in the twentieth century has not only increased in terms of manpower, organization, apparatus, research output, and economic support, but it has also progressed scientifically, that is, it has produced much new knowledge about nature.[80]
>
> ...
>
> Physics will undoubtedly continue to develop and make many interesting discoveries in the new century.[81]

In what way does Kragh's historical account contribute to progress in physics? The account given partly is about

progress ... as conspicuous a feature of twentieth-century physics as has growth.[82]

But, there is no analysis, no attempt to explain the "important parts of the development."[83] All the same, the book is a nuanced story-line written by an experienced historian in physics. The account given is

condensed and selective.[84]

It gives chronological orderings of various developments in physics, and points also to general aspects of the whole community, in the years 1895–1995.[85]

---

[79] Kragh [1999], Superstring Theory, pp. 415–419.
[80] Kragh [1999], p. 443.
[81] Kragh [1999], p. 450.
[82] Kragh [1999], p. 444.
[83] See note 76.
[84] See note 74.
[85] See note 76. Kragh's work might well eventually be recycled.

3.2.5 *The Oxford Handbook of the History of Physics*[86]

The *Oxford Handbook of the History of Physics* includes a wide variety of articles.

> The *Handbook* brings together chapters on key aspects of physics from the seventeenth century to the present day.[87]

Discussions include the gradual emergence of physics as distinct from both natural philosophy and chemistry; other factors in society such as religious, social, political, military, industry and technology; patronage; physics and instrument making; physics and medicine, meteorology and education; imagery, diagrams and symbolism used by different physicists at different times; leading questions of particular times; developments in mathematical modeling. Some chapters tell the stories of the emergence of statistical mechanics, quantum mechanics, relativity, and recent changes in technology[88]. Chapter 24 looks to origins of the names 'classical' and 'modern physics.'[89] And, Chapter 26 is on the origins of the modern theory of relativity[90]. In his article on relativity, Kennefick recalls Einstein's 1905 paper On the Electrodynamics of Moving Bodies,[91] and draws special attention to an increasingly evident division of labor in the community:

> The complexity of the mathematics required to handle these issues was encouraging the creation of a new breed of scientist, the theoretical physicist, who specialized in mathematically complex and difficult calculations.[92]

---

[86] Buchwald and Fox [2013].

[87] Buchwald and Fox [2013]. In *The Oxford Handbook*, ch. 2 on Galileo, ch. 3 on Cartesian Physics, and chapters 5 and 6 on Newton's work, each emphasize interpretation of the meanings of individual authors. The other chapters of *The Oxford Handbook* focus on historical developments.

[88] Shinn [2013].

[89] Gooday and Mitchell [2013].

[90] Kennefick [2013].

[91] Available in Stachel, Cassidy, Renn, and Schulmann, R. [1990], ch. 23.

[92] Kennefick [2013], p. 790.

Chapter 29[93] of the *The Oxford Handbook* [94] is by Helge Kragh. As in his book *Quantum Generations*[95], here too, there is no analysis[96]. The article outlines sequences of names, results and dates. The focus of the article is twentieth-century cosmology, which

> has far from developed smoothly or linearly, but in spite of many mistakes and blind alleys ... has progressed remarkably.[97]

Kragh's article begins by recalling Doppler's work; it lists names of follow-up contributors; it goes on to include summary paragraphs of Einsteinian Cosmology, with a penultimate section on recent (*circa* 2012) observations of cosmic microwave background,

> greatly improved in quantity and quality with the launching in 1989 of the COBE satellite, carrying instruments specially designed to measure background radiation over a wide range of wavelengths.[98]

The last section of Kragh's paper is Post-Script: Multiverse Speculations.[99] The article ends by drawing attention to an ongoing controversy:

> Perhaps the most controversial of the modern cosmological hypotheses is the idea of numerous separate universes, or what is known as the 'multiverse' – a term first used in a scientific context in 1998.
>
> ...
>
> Remarkably, many cosmologists and theoretical physicists have become convinced that our universe is just one out of perhaps 10^{500} universes. ... Almost all physicists agree that a scientific theory has to speak about nature in the sense that it must be

---

[93] Kragh [2013].
[94] Buchwald and Fox [2013].
[95] Kragh [1999]. See Section 3.2.4 .
[96] See note 102.
[97] Kragh [2013], p. 892.
[98] Kragh [2013], Physics and Cosmology, p. 915.
[99] Kragh [2013], Multiverse Speculations, sec. 29.10, pp. 918–919.

empirically testable, but they do not always agree what 'testability' means or how important this criterion is relative to other criteria.[100]

3.3   Some Non-Trivial Problems to Resolve

Scholars referred to in Section 3.2 are well recognized for publications in their respective areas. In this section, I draw attention to a few problems in historical method. I do this not because the problems are unique to the works of these authors, but because the works discussed here provide evidence of present-day trends.

A main feature of *Quantum Generations* is seen in the following:

> With a few exceptions, I have avoided equations, and although the book presupposes some knowledge of physics, it is written mainly on an elementary level.[101]

What is it that is communicated? As already quoted, Kragh

> described, rather than analyzed, important parts of the development of physics between 1895 and 1995.[102]

The book is intended to help a reader get a

> big picture – to evaluate the revolutionary changes and make comparisons over the course of a century.[103]

But, how does one evaluate the story of physics without understanding physics, which of course, includes understanding equations? Without understanding the physics, are we not limited to merely nominal orderings, such as names, dates and places? In fact, the results of the book mainly are nominal. And, the concluding Part Four: A Look Back[104] is a nominal summary of previous nominal accounts. The book, then, is less a description of developments in physics[105] than it is a potentially helpful

---

[100] Kragh [2013], Physics and Cosmology, pp. 918–919.
[101] Kragh [1999], p. xiii.
[102] Kragh [1999], p. xiv.
[103] Kragh [1999], p. xiv.
[104] Kragh [1999], Part Four: Look Back, pp. 425–451.
[105] See note 102.

historical index, indicating points of entry by which one might obtain data on developments in physics[106].

In Part Four of *Quantum Generations*, three kinds of progress are described:

extensive[107];

opening new windows through which nature's secrets can be studied[108]; and

a third kind of progress, which is essentially theoretical, is the introduction of new frameworks and principles that reorganize and give new meaning to already obtained knowledge.[109]

On what grounds is this list determined? Is the list complete? How do the three kinds of progress Kragh points to bear out in the field? Recall, as quoted above:

Not only are many of the world's physicists occupied with applied aspects of their science, and have been so during most of the century, but it is also largely through the technological applications that physics has become a major force of societal change.[110]

But, Kragh's work is part of physics too! Does it, however, fit within the three kinds of progress that he indicates?

As noted in Section 3.2.2, the paper by Jackson and Okun is a record of names, dates, formulas and experiments in the early history of gauge theories. A different issue is raised, however, in the fourth section of their paper, namely, the physical meaning of gauge invariance:

As has been emphasized above, gauge invariance is a manifestation of nonobservability of A^μ. However, integrals ... are observable when they are taken over a closed path, as in the Aharonov-Bohm effect (Aharonov and Bohm, 1959). The loop integral of the vector potential there can be converted by Stokes's theorem

---

[106] Key data is obtained by adverting to one's experience in one's own understanding, and development in understanding. I am referring to a needed but not yet implemented generalized empirical method. See ch. 6.

[107] Kragh [1999], p. 443.

[108] Kragh [1999], p. 444.

[109] Kragh [1999], p. 444.

[110] See note 74.

into the magnetic flux through the loop, showing that the result is expressible in terms of the magnetic field, albeit in a nonlocal manner. It is a matter of choice whether one wishes to stress the field or the potential, but the local vector potential is not an observable.[111]

In what ways might these observations help us explain the historical development of gauge theory? For example, what is the historical significance of a discovery that, depending on one's choice, is about field equations or potential functions? Gauge theory evolved in an ongoing effort to account for real things called elementary particles. That historical development included attempts to understand the real significance of integrals, defining relations and boundary conditions. In particular, while the historical development has included a growing proficiency in observational methods and computational techniques, it also has included an ongoing struggle to understand the real significance of different potential functions[112].

Let us now look again to O'Raifeartaigh's book. As discussed in Section 3.2.1, above, *The Dawning of Gauge Theory* invites the reader to enter into details in the evolution of gauge theories. The book includes new descriptions of details in developments, and ends with questions and pointers toward future progress. If one rises to the challenge of reading the book with understanding, one will be re-enacting the historical evolution and that way one also will be obtaining data on development in physics.[113] Implicitly, then, the invitation of the book goes well beyond a merely nominal account. However, as is normal at this time in history, *The Dawning of Gauge Theory* does not explicitly invite the reader to a dawning in self-attention[114].

Not unlike Kragh's book, Smolin's book *The Trouble with Physics* also is intended for

experts and lay readers alike.[115]

---

[111] Jackson and Okun [2001], p. 676.

[112] For advanced and future readers, one may note here that a major challenge is to make progress toward determining metaphysical equivalence (Lonergan [1992], Secs. 16.3.3, 16.3.4), and so to be luminous in differences between primary relations and secondary determinations in contemporary physics (Secs. 16.1 and 16.2).

[113] See note 106.

[114] See note 106.

[115] Smolin [2006], p. xviii.

Again, though, for lay readers, the book will be largely inaccessible. Why do I say that? Smolin communicates his reflections on a complex weave of advanced developments in modern physics, namely, 20th and (early) 21st century string theory. Part II of Smolin's book does provide something of an aerial view of part of what was going on in physics, over a time period of about 100 years – but meaningful only if one knows what the words mean! And, as one finds from experience, there is no shortcut or substitute: one gets to know what the words of modern physics mean by learning modern physics. As physics graduate students know, that is an exciting but difficult climb that requires considerable background and usually takes years of sustained effort.

Similar comments apply to some of the articles in *The Oxford Handbook of the History of Physics*. To be sure, my few and brief comments do not do justice to the scholarship there, which in turn is only a small sample from the literature. My purpose here, though, is to invite interest in difficulties in contemporary methods. And so, I think that, for now, enough details have been given, at least in order for us to get a few first impressions of the present-day situation in historical studies in physics.

Two issues of particular significance are: In historical work, we need to distinguish between understanding how to use the words of physics and understanding what the words of physics mean. Also evidently lacking is a common heuristics of development, within which historians could make progress toward explaining transitions. For instance, words like 'force' and 'charge' remained in use, but took on new colorful meanings[116] in new explanatory contexts.

3.4 Progress-Oriented Historical Studies

Section 3.3 invites attention to problems in historical method in physics, evidenced in works cited in this book. How, then, are we to study the history of physics in ways that (a) move toward resolving some of the

---

[116] I am, of course, referring to 'colour charge.'

problems noted in the previous section; and, at the same time, (b) contribute to progress in physics?

A few observations are possible here. As O'Raifeartaigh's book makes explicit, and Smolin's book implicitly requires, the task of historical studies needs the historian and the reader to understand the appropriate physics. But, explaining the history of physics is more than understanding various physics of various times. The problem includes understanding something of developments that led up to the physics of a time. One needs to be familiar with prior experimental methods, technologies and questions; have representative examples of experimental data of past and present physics; if one is studying, say, the history of gauge theory, or string theory, one needs to have some understanding of various field theories, or string theories, that were part of the historical development leading up to the Standard Model, and current string theories. In fact, even if one ultimately is an advocate of string theories, if one is to understand the historical development of string theory, one also needs to also know the Standard Model that string theory finds inadequate and seeks to replace.

Moreover, one needs at least some experience for oneself in: explaining particular trajectories by appealing to actual boundary conditions, lengths, measurements and terms defined within older geometries; as well as doing so in the more recent gauge theory geometry of the present-day Standard Model. But, still, that is not enough. For, if one only has experience explaining trajectories in different geometries, one only has data on moments in the historical development. One also needs to identify key transitions between different physics and geometries in one's own experience. In other words, if one is going to study historical development of physics, one needs data on that development. And, data on development in understanding is obtained within one's own understanding, by re-enacting key moments and transitions in historical developments in observing and understanding.

These observations, however, are merely hintings of aspects of what Lonergan called *generalized empirical method*.[117] But, having now tuned

---

[117] See note 106.

somewhat to sources, we can segue to Lonergan's dense but more complete description of historical understanding:

> The history of any particular discipline is in fact the history of its development. But this development, which would be the theme of a history, is not something simple and straightforward but something which occurred in a long series of various steps, errors, detours, and corrections. Now, as one studies this movement one learns about this developmental process and so one now possesses within oneself an instance of that development which took place perhaps over several centuries. This can happen only if the person understands both his or her subject and the way he or she learned about it. Only then will he or she understand which elements in the historical developmental process had to be understood before the others, which ones made for progress in understanding and which held back, which elements belong to the particular science and which do not, and which elements contain errors. Only then will he or she be able to tell at what point in the history of the subject there emerged new visions of the whole and when the first true system occurred, and when the transition took place from an earlier to a later systematic ordering, which systematization was simply an expansion of the former and which was radically new; what progressive transformation the whole subject underwent; how everything that was explained by the old systematization is now explained by the new one, along with many other things that the old one did not explain – the advances in physics, for example, by Einstein and Max Planck. Then and only then will he or she be able to understand what factors favored progress, what hindered it, and why, and so forth.

> Clearly, therefore, the historian of any discipline has to have a thorough knowledge and understanding of the whole subject. And it is not enough that he or she understand it in anyway at all, but he or she must have a systematic understanding of it. For that precept, when applied in history, means that successive systems which have been progressively developed over a period of time have to be understood. This systematic understanding of a development ought to make use of an analogy with the development that takes place in the mind of the investigator who learns about the subject, and this interior development within the mind of the investigator ought to parallel the historical process by which the science itself developed.[118]

---

[118] Lonergan [2013], pp. 175–177. Lonergan [2013] is taken from *De Intellectu et Methodo*, 1959, lecture notes for students. See Lonergan [2013], p. 3, fn. 1. Volume 23 of the *Collected Works* has the original Latin on even-numbered pages, and corresponding English translation in the subsequent odd-numbered page. The word 'system' is not imposing Kuhnian idealized structure. Much as the biologist who studies growth and development, the problem is empirical.

This is an amazing two paragraphs and will be further discussed in Chapter 7. Already, though, the quotation can help us sense something of a much larger significance of Lonergan's heuristics of *explanatory interpretation*, discussed briefly in Chapter 2[119]. But, as also discussed in Chapter 2, interpretation, as such, of Lonergan's work is a future possibility.[120] We can, though, take help from the quotation. For, whatever the genius Lonergan intended, the quotation can serve as a descriptive pointer.[121]

Immediately, readers may note that there are many histories written about many different kinds of development. In this book alone, I have cited historical works regarding: the development of gauge theories; the history of string theory; and attempts at broader tellings, such as in the book and paper by Kragh. But, there are also histories written about other kinds of development in physics, and related areas.[122] Moreover, written histories can differ greatly, even when they are about the same development. They also can differ in areas of focus; in the times and places in which they were written; and in basic assumptions and views of historians[123].

---

[119] Section in 2.6. But, lift to the community task that is eightfold.

[120] If, for example, we take the meaning of the quotation above as obvious, then we fall into the trap of attempting to substitute nominal understanding for explanatory understanding.

[121] See ch. 6. This is normal in science. Descriptive results often are taken to be provisionally true, awaiting explanation.

[122] The history of the sciences literature is large, so in order to illustrate, I only mention a few areas. There are, for example: written histories about biophysics; histories of biophysics contribute to histories of biochemistry; and there are written histories of neuroscience which figure into histories of medical physics and technology. Indeed (without implying a reductionism), recent histories of neuroscience regard progress that makes it possible now to partly distinguish states of consciousness (e.g. dreaming and wakefulness) as well as modes of insight, deliberation and decision, in terms of the biophysics and biochemistry of the brain (what in neuroscience are called *neural correlates*). There are histories written about developments in pedagogy (e.g., histories of pedagogy in physics (Meltzer and Otero [2015])); histories regarding developments in methods of physics and other sciences; histories of the physics of archeology; histories of views in astrophysics, cosmology, and emergence, complexity, ultimate reality, Space, Time and God (Kragh [2004]).

[123] Compare, e.g., Smolin [2006], Part II, with Schwarz [2012]. Smolin [2006], Part II highlights both potential strengths and potential weaknesses in the development of string

Following on these observations, and to conclude this chapter, I draw attention to two sets of issues. The first regards the possibility of a mature scientific historical studies. Historical studies is, of course, a long established and developing mode of scholarship. But, as touched on briefly in Section 3.3, such scholarship struggles without a shared heuristics of historical understanding that would contribute to progress.

Let us call the (not-yet-obtained) progress-oriented historical studies *functional history*. Even though not-yet-differentiated, some elements of the needed heuristics can be witnessed in histories written (for example, those cited in this chapter). In some cases, elements needed are conspicuous by their absence, for example, as in a lack of control of meaning. Functional history will reach new understandings of sequences; data will need to include the historian's own development (in other words, functional history will be an achievement within *generalized empirical method*[124]). The task of functional history will include: an effort to identify genera and species of transitions in sequences – advances, breakthroughs, positive shifts, but also prolongations of, or newly generated confusions; an effort to identify progress as well as deviation from progress (relative to one's – the historian's – view of progress and decline[125]). The task will not merely be interest in the past, "history for history's sake," but "history

---

theory. By contrast, Schwarz [2012] gives a fundamentally positive slant on the same history: "This lecture presents a brief overview of the early history of string theory and supersymmetry. It describes how the S-matrix theory program for understanding the strong nuclear force evolved into superstring theory, which is a promising framework for constructing a unified quantum theory of all forces including gravity. The period covered begins with S-matrix theory in the mid 1960s and ends with the widespread acceptance of superstring theory in the mid 1980s. Further details and additional references can be found in Schwarz (2007)" (Schwarz [2012], p. 1). In the Epilogue of the same paper: "For many years string theory was considered to be a radical alternative to quantum field theory. However, in recent times – long after the period covered by this lecture – dualities relating string theory and quantum field theory were discovered. In view of these dualities, my current opinion is that string theory is best regarded as the logical completion of quantum field theory, and therefore it is not radical at all. There is still much that remains to be understood, but I am convinced that we are on the right track and making very good progress" (Schwarz [2012], p. 10).

[124] See note 106 and 117.

[125] Of course, an historian's views can and do change.

for our future's sake."[126] However, to go beyond these vague anticipations will require progress toward and in implementation.

The second set of issues acknowledges further kinds of question that arise. A written history tells only part of our whole story. How do we distinguish, evaluate and compare different written histories? How do we reach for a whole story? Such questions arise and, for example, are implicit in basic views of individual historians. These issues, then, are indications of a further task which will be discussed in Chapter 4.

3.5 References

Barrow, J. D., Davies, C. W., and Harper, C. L. Jr. eds. (2004) *Science and Ultimate Reality: Quantum Theory, Cosmology, and Complexity* (Cambridge University Press, Cambridge).

Bérard, A., Mohrbach, H., Lages, J., Gosselin, P. , Grandati, Y. , Boumrar, H. , Ménas, F. (2007) From Feynman ['s] proof of [the] Maxwell equations to non-commutative quantum mechanics, in *Third Feynman Festival, JPCS*, 70 (IOP Publishing, Bristol, UK).

Buchwald, J. Z. and Fox, R., eds. (2013) *The Oxford Handbook of the History of Physics* (Oxford University Press, Oxford).

Dyson. F. J. (1990) Feynman's Proof of Maxwell's Equations, *Am. J. Phys.*, 58 (3), pp. 209 – 211.

Gibney, E. (2016) Hopes for revolutionary new LHC particle dashed, *Nature – Internat. Weekly J. Sci.*, vol. 536, issue 7615, August, pp. 133–134, doi:10.1038/nature.2016.20376.

---

[126] The lean is familiar in, for example, veterinary science. The veterinarian may need to learn about a dog's past (nutrition, previous illnesses, activities, environment, and so on). The motivation, though, is to help the animal overcome present difficulties, and do better in the future. That future lean, caring for the community, also is evident in the works of some historians. For example, we find the following, in an article on the history of physics education: "The discussion should help physics educators ... evaluate the effectiveness of various education transformation efforts of the past, so as better to determine what reform methods might have the greatest chances of success in the future" (Meltzer and Otero [2015], p. 447). Within a functional division of labor, there will be the forward functional specialties. As will be discussed in chapters below, results of functional history will contribute to functional doctrines (ch. 6). However, within functional cycling, there also will be the work of functional dialectics (ch. 4) and functional foundations (ch. 5).

Goenner, H. F. M (2004) On the History of Unified Field Theories, *Living Rev. Relat.*, 7 (2004).

Gooday, G. and Mitchell, D. J. (2013) Rethinking 'Classical Physics,' ch. 24 in Buchwald, J. Z. and Fox, R. eds., *The Oxford Handbook of the History of Physics* (Oxford University Press, Oxford), pp. 721–764.

Greene, B. R. (2003) *The Elegant Universe. Superstrings, Hidden Dimensions, and the Quest for the Ultimate Theory* (W. W. Norton, New York).

Jackson, J. D. and Okun, L. B. (2001) Historical roots of gauge invariance, *Rev. Mod. Phys.*, vol. 73, July, pp. 663–680.

Kennefick, D. (2013) Three and a Half Principles: The Origins of Modern Relativity Theory, ch. 26 in, Buchwald, J. Z. and Fox, R. eds., *The Oxford Handbook of the History of Physics* (Oxford University Press, Oxford), pp. 789–813.

Kragh, H. (1999) *Quantum Generations. A History of Physics in the Twentieth Century* (Princeton University Press, Princeton).

Kragh, H. (2004) *Matter and Spirit in the Universe: Scientific and Religious Preludes to Modern Cosmology*, vol. 3 of *History of Modern Physical Sciences* (Imperial College Press, London).

Kragh, H. (2013) Physics and Cosmology, ch. 29 in, Buchwald, J. Z. and Fox, R. eds., *The Oxford Handbook of the History of Physics* (Oxford: Oxford University Press, 2013), pp. 892–922.

Lonergan, B. (1992) *Insight: A Study of Human Understanding, Collected Works of Bernard Lonergan*, vol. 3, eds. Frederick E. Crowe and R. M Doran (University of Toronto Press, Toronto).

Lonergan, B. (2013) *Early Works on Theological Method 2*, trans. Michael G. Shields, eds., Robert M. Doran and H. Daniel Monsour, *Collected Works of Bernard Lonergan*, vol. 23 (University of Toronto Press, Toronto).

Meltzer, D. E. and Otero, V. K. (2015) A brief history of physics education in the United States, *Am. J. Phys.* 83 (5), May, pp. 447–458.

O'Raifeartaigh, L. (1997) *The Dawning of Gauge Theory* (Princeton University Press, Princeton).

O'Raifeartaigh, L. and Straumann, N. (2000) Gauge theory: Historical origins and some modern developments, *Rev. Mod. Phys.*, vol. 72, no. 1, January.

Pessoa, O. Jr. (2001) Counterfactual Histories: The Beginning of Quantum Physics, *Philos. Sci.*, vol. 68, no. 3, *Supplement: Proceedings of the 2000 Biennial Meeting of the Philosophy of Science Association. Part I: Contributed Papers*, Sept., pp. S519–S530.

Pombo, C. (2009) A New Comment on Dyson's Exposition of Feynman's Proof of Maxwell Equations, *AIP Conf. Proc.*, 1101 (2009), p. 363.

Schwarz, J. H. (2012) The Early History of String Theory and Supersymmetry, https://arxiv.org/pdf/1201.0981v1.pdf, Jan. 4.

Shinn, T. (2013) The Silicon Tide: Relations between Things Epistemic and Things of Function in the Semiconductor World, ch. 28 in, Buchwald, J. Z. and Fox, R. eds.,

The Oxford Handbook of the History of Physics (Oxford University Press, Oxford), pp. 860–891.

Smolin, L. (2006) A Brief History of String Theory, Part II in, *The Trouble with Physics, The Rise of String Theory, the Fall of a Science, and What Comes Next* (Houghton Mifflin, Boston).

Stachel, J., Cassidy, D. C., Renn, J. and Schulmann, R. (1990) *The Collected Papers of Albert Einstein, Volume 2: The Swiss Years: Writings, 1900-1909* (Princeton University Press, Princeton).

Stachel, J., Cassidy, D. C. and Schulmann, R., eds. (1997) *The Collected Papers of Albert Einstein*, vol. 7 (Princeton University Press, Princeton: Princeton University Press).

Straumann, N. (1996) Early History of Gauge Theories and Weak Interactions, Invited talk at the *PSI Summer School on Physics with Neutrinos* (*Physics with neutrinos*, Zuoz, Swizterland), pp. 153–180.

Taylor, J.C., ed. (2001) *Gauge Theories in the Twentieth Century* (Imperial College Press, London).

Chapter 4

Functional Dialectics

4.1 Introduction

Smolin tells us that one of the reasons he wrote *The Trouble with Physics*[1] is that he was

> concerned about a trend in which only one direction of research is well supported while other promising approaches are starved[2].

Part II of his book is A Brief History of String Theory[3], in the last chapter of which[4] Smolin offers an

> assessment of string theory as a proposal for a scientific theory.[5]

He suggests that

> string theory is certainly among the directions that deserve more investigation. ... But ... string theory has not be successful enough on any level to justify putting nearly all our eggs in its basket.[6]

However, it is also true that there are different histories written about string theory, written by authors with different points of view, with different and sometimes opposing assessments of whether or not string theory has been making progress as a scientific theory of elementary

---

[1] Smolin [2006].
[2] Smolin [2006], p. xxiii.
[3] Smolin [2006], Part II, A Brief History of String Theory, pp. 101–199.
[4] Smolin [2006], ch. 12, pp. 177–199.
[5] Smolin [2006], p. 177.
[6] Smolin [2006], p. 199.

particles[7]. At this stage there are two main views on the issue. More traditional physicists require that theory be checked in experimental data. Proponents of string theory, by contrast, do not insist on such data but appeal, instead, to mathematical beauty and invoke contemporary philosophical arguments[8].

How can physics make progress toward resolving these basic and fundamental difficulties? Must we resign ourselves to not only a landscape of string theories, but also an expanding landscape of philosophical debates? Or, might there be a way to make progress toward reaching consensus about fundamental directions?

No doubt, an

assessment of string theory as a proposal for a scientific theory[9]

partly depends on one's view of what constitutes a scientific theory. So, the problem, at least in part, involves philosophy of science. But, even so, we will need a special kind of practical philosophy of science. Among other things, it will need to be able to help physics account for and compare detailed results from: string theory; the Standard Model; and results being obtained in contemporary experimental physics.

Recognizing that philosophy of science is needed raises further problems. As Crull observes,

(d)ebates in particle physics have ... contributed new material for foundational questions relating to realism, the identity or indiscernibility of objects, and even

---

[7] See note 123 in ch. 3. See also, e.g., Cappelli, Castellani, Colomo and Di Vecchia [2012] and Rickles [2014].

[8] In more detail: "'What is the role of theoretical physics?' There are two extreme views as to what this role should be. One is that the role of theory is to be closely tied to the experimental and phenomenological field, help experimentalists interpret their experiments and distinguish signal from noise. The other attitude is that the goal of theoretical physics is to achieve a higher level of understanding. In order to achieve such understanding one might focus on the solution of well-defined mathematical models consistent with general principles, regardless of whether those models are real of not. Indeed, how much value do we assign to simplicity and mathematical elegance? That is what the second group usually cares about" (Gross [2005], pp. 5908–5909).

[9] Smolin [2006], p. 177.

(what Crull calls)[10] The Ultimate Meta-Question of how disciplines like metaphysics, philosophy of physics and physics ought to interact with, inform and edify one another. ... Suffice it to say that signs point to the mutual, if slow, dissolution of both the anti-metaphysics club in philosophy of physics and the physics-phobic armchair club in metaphysics. Legitimate progress on this front is eagerly awaited.[11]

So far, though, contemporary philosophy of physics has made little progress in helping physics resolve controversies or make definitive progress in sorting out basic issues regarding string theory and the present Standard Model. It is not that there is any shortage of publications. But, ongoing philosophical work notwithstanding, contemporary philosophy of physics is not showing signs of reaching consensus about how philosophy of physics and physics are related.

There is, however, an emerging consensus that physics and philosophy of physics are somehow mutually dependent:

The boundary between physics and philosophy of physics is blurry[12].

In a similar way:

(T)here is no sharp line between philosophy of physics and physics itself.[13]

The lack of consensus is poignant when contemporary philosophy of physics attempts to answer questions about particular issues. For instance, whether one is looking to the Standard Model or to string theory, both involve quantum theory. But,

(s)ince the 1960's, the physics community has witnessed a revival of the debates about the interpretation of quantum theory that raged among the theory's founding fathers.[14]

---

[10] Parenthetic phrase inserted.

[11] Crull [2013], pp. 780–781.

[12] Kuhlmann and Pietsch [2012], Thesis (ii), pp. 210–213. See also, Rickles [2008], p. 6.

[13] Butterfield and Earman [2007], p. xviii.

[14] Butterfield and Earman [2007], p. xv. In this chapter, the word 'interpretation' is from contemporary philosophy of physics and science, and not be confused with 'functional interpretation,' a not-yet-operative functional specialty.

In her review article, Crull recalls some of the more familiar interpretations of quantum physics:

> hidden variables,..., Copenhagen, ..., wave-function collapse, ..., alternate hidden variables theories in terms of modalities, ..., ontology of collapse theories has been a hot topic as well.[15]

As Crull also points out, however,

> (t)he list of interpretations of quantum mechanics is of course more extensive than those catalogued above. Indeed, recent years have seen a rise in alternate perspectives ranging from ontic commitment to the quantum state (e.g. wave-function or state realisms of the sort described in the collection by Ney and Albert 2013) to proposals wherein quantum probabilities are understood qua propensities (summarized in Suárez 2007). In 2003, Fuchs suggested that quantum mechanics be understood almost exclusively in terms of information, and this information-theoretic approach has met with growing enthusiasm in both physics and philosophy circles.[16]

What are we to do? Evidently, the problem is not mere logic. Comparably skilled scholars intelligently argue for mutually incompatible views. There are also obvious differences in method. Some scholars appeal to combinations of mathematics and the results of experimental physics. Analytic philosophy tends to emphasize precision in logic and "text-based methodologies."[17] String theorists look more to mathematical constructs, with a conviction in the validity of their approach.

What we find, then, is that different scholars appeal to different criteria, and that these criteria regard, for example, what is scientific, or not; what is plausible, or not; and what is real, or not. At the same time, current methods do not ask that scholars bring such criteria out into the open. And, again, as history shows, this allows for, and even promotes, more or less endless debate. Scholars from different camps talk at cross-purposes, often use the same words but with different meanings, use comparable logic, but

---

[15] Crull [2013], pp. 772–773.
[16] Crull [2013], p. 773. See also, Butterfield and Earman [2007], p. xv: "Since the 1990's, the burgeoning fields of quantum information and computation have grown out of the interpretative debates, especially the analysis of quantum non-locality."
[17] Kuhlmann and Pietsch [2012], Thesis (v), pp. 211.

implicitly appeal to different basic criteria, indeed, basic criteria which current methods do not discern.

Might it not be helpful if at least some subgroup of scholars were to make basic criteria explicit? Might it not be advantageous to make an effort to bring one's basic criteria into the open, to be as explicit as possible in, and about, one's stand on, for example, what is scientific, or not; what is plausible, or not; and what is real, or not? This would not be for the purposes of winning a debate. It would be as a contribution to scientific collaboration. It would be so that a subgroup of scholars might collaboratively compare, be challenged to make adjustments, and, as much as possible, move toward consensus about basic criteria and views.

Yet, this may all sound somewhat implausible. What criteria are more basic than those already provided in contemporary philosophy of physics? What could be better than what is already being done, that is, stipulate hypotheses and premises and work out derived terms in a logical system? So, what might it look like to make one's criteria of reality explicit? How might one do so in ways that would help the scientific community move toward consensus and be better able to handle specific problems such as: the physical significance of various quanta, the physical significance of gauge invariance, or the nature of the Higgs field?

We do not yet know how all of this might be possible. We are not there yet. For, what I am hinting at in these paragraphs are only a few elements of a not-yet-operative complex task that Lonergan called *functional dialectics*. While the heuristics of that task were mapped out brilliantly by him (*Method* 250[18]), functional collaboration is not yet operative and, in particular, nor is functional dialectics. But, this need not be cause for concern here. For, remember that the purpose of this book is more elementary, namely, to cultivate interest in helping promote the emergence of that future eightfold division of labor. Learning how to collaborate functionally will be a future work in progress.

Like the rest of this book, the purpose of this chapter, then, is modest. Specifically, this chapter is to invite the reader to take a minimal stand regarding experience in physics. This is so basic that, eventually, it can

---

[18] Lonergan [1975], Bottom two lines of p. 249 and all of p. 250. For future reference, named *Method* 250.

lead us to beginnings in an improved empirical method and a new control of meaning[19]. But, that development will come later in history. To the immediate purpose of this chapter, taking a minimal stand (whatever it is) about one's own experience, and then also making a beginning in recognizing the potential advantage of making one's stand explicit, can help intimate something of a fourth task called *functional dialectics*.

Some readers, though, already may be inclined to disagree that taking such a stand is either feasible or worthwhile. And yet, does not expressing such disagreement confirm the feasibility and the need? In that disagreement, are you not taking a stand? Does your claim cohere with what you do when you are doing physics? Indeed, does your claim cohere with making your claim? Does your claim make a positive contribution to the community? As the reader may find, attempting to answer any of these questions further contributes to revealing something of the fourth task.

Section 4.2 has a focus topic, the problem of what is often called "making sense of probabilities in physics." The main purpose, though, is to help reveal the need for attending to experience in doing physics. This is done by working on a particular problem and by taking steps toward *illuminating experience*. As the reader will find out, the word 'illuminating' here is meant both as an adjective and a transitive verb. Where Section 4.2 is to obtain hintings of a fourth functional specialty, Section 4.3 is a pointing to the complex future achievement precisely outlined by Lonergan in *Method* 250[20].

4.2   Making Sense of Probabilities in Physics by Illuminating Experience

As mentioned in Section 4.1, the main objective of this section is methodological. However, the focus topic is probabilities[21] in physics. And, of course, many questions arise. For instance,

---

[19] Generalized empirical method: see Chapter 6.

[20] See note 18.

[21] As noted by Uffink, "historical development of the theory of probability has been rather well documented and studied. The first main text in this field is Jakob Bernoulli's *Ars Conjectandi* (published posthumously in 1713). Indeed, it is the source in which the term

what are probabilities and how can we explain the meaning of probabilistic statements? How can one justify physical claims that involve probabilities? Finally, can we draw metaphysical conclusions from the abundance of probabilities in physics? For example, can we infer that we live in an inherently chancy or indeterministic world?

Although these are distinct questions, they are connected and cannot be addressed in isolation. For instance, an account of the meaning of probabilistic statements would clearly be objectionable if it did not yield a plausible epistemology of probabilities. Further, the metaphysical lessons that we may wish to draw from the abundance of probabilistic claims hinge on the meaning of 'probability.' Hence, our three questions set one major task, viz. *to make sense of probabilities in physics.*[22]

In the attempt to make sense of probabilities in physics,

a dichotomy of two broad groups of views has emerged: ... between objectivist and subjectivist views.

...

According to *objectivist*[23] views, probabilistic statements state matters of fact. That is, they have truth-conditions that are fully mind-independent and refer to frequencies of events or to propensities in the world. We can then use descriptions of truth-conditions to form simple slogans such as 'probabilities are frequencies.' *Subjectivist* views, by contrast, take probabilistic statements to express degrees of belief. Thus, a fitting slogan is 'probabilities express credences.' Of course, we cannot assume that all probabilistic statements from physics and elsewhere are to be interpreted along the same lines. Hence, pluralist accounts of probabilities suggest different interpretations in different domains of discourse.[24]

A similar sketch of two main groupings of views is given by Frigg, but includes subdivisions:

'probability,' in its modern quantitative sense, is coined" (Uffink [2011], p. 26). But, while Bernoulli (1700–1782) produced the first main text, the story goes back further still. For, "the true founders of the mathematical theory of probability were Pascal (1623–1662) and Fermat (circa 1601/7-1665), who developed the fundamental principles of the subject in an intensely interesting correspondence during the year 1654" (Bell [1937], p. 86). The correspondence between Fermat and Pascal is available in Tannery and Henry [1904].

[22] Beisbart and Hartmann [2011], Introduction, p. 1.

[23] Italics in source text.

[24] Beisbart and Hartmann [2011], p. 4.

Approaches to probability can be divided into two broad groups. First, *epistemic approaches* take probabilities to be measures for degrees of belief. Those who subscribe to an *objective epistemic theory* take probabilities to be degrees of rational belief, whereby 'rational' is understood to imply that given the same evidence all rational agents have the same degree of belief in any proposition. This is denied by those who hold a *subjective epistemic approach*, regarding probabilities as subjective degrees of belief that can differ between persons even if they are presented with the same body of evidence. Second, *ontic approaches* take probabilities to be part of the 'furniture of the world.' On the *frequency approach*, probabilities are long run frequencies of certain events. On the *propensity theory*, probabilities are tendencies or dispositions inherent in objects or situations. The *Humean best systems approach* – introduced by Lewis (1986) – views probability as defined by the probabilistic laws that are part of that set of laws which strike the best balance between simplicity, strength and fit.[25]

For a nudge toward attending to experience, I give the following question: In order to interpret probabilities in physics, might it make sense to attend to what one does when obtaining or reaching probabilities when doing physics? To be sure, a growing plurality of

accounts of probabilities suggest different interpretations in different domains of discourse.[26]

But, in order that interpretations be of probabilities in physics then, surely, it is to doing physics that we must go.

To which probabilities in physics shall we go? In order for results to be up-to-date, we will eventually need to consider events, occurrences, state vectors and probability functions defined in the Standard Model; statistical distributions and other experimental results from cyclotrons like the Large Hadron Collider; as well as results in contemporary statistical mechanics and thermodynamics, to name but a few of many zones in modern physics where numerical probabilities feature in both theory and application. However, there are elementary insights needed before moving into those advanced contexts. And, in fact, we can get those elementary insights by looking to the much-studied problem of coin-tossing, a problem that continues to be *fascinating*[27].

---

[25] Frigg [2008], pp. 114–115.
[26] Beisbart and Hartmann [2011], p. 4.
[27] Diaconis, Holmes and Montgomery [2007], p. 233.

Even though it is an old and elementary problem, three authors from Stanford University and UC Santa Cruz recently got new results. The context of their work is stochastic processes and dynamical systems. They did a new study of coin tossing, using a well-engineered coin tossing machine.[28] For what they called

natural flips[29],

they found that

the chance of coming up as started is about .51.[30]

Although, they also point out that in order

(t)o detect the departures of the order of magnitude we have found would require 250.000 tosses. The classical assumptions of independence with probability 1/2 are pretty solid.[31]

Let's now make a small beginning in comparing results of Diaconis *et al.* with the interpretations of probability outlined above[32], namely: *objectivist, ontic*; and *subjectivist, epistemic*. In fact, I will only discuss a few aspects of these views. That will be enough, though, for us to make a beginning in teasing out certain methodological issues, which is the main purpose for this section.

I start by looking to the *objectivist* view. As found within the quotation above,

(a)ccording to *objectivist* views, probabilistic statements state matters of fact. That is, they have truth-conditions that are fully mind-independent and refer to

---

[28] Diaconis, Holmes and Montgomery [2007]. "(T)he discussion … highlights the true difficulty of carefully studying random phenomena. If we have this much trouble analyzing a common coin toss, the reader can imagine the difficulty we have with interpreting typical stochastic assumptions in an econometric analysis" (Diaoconis, Holmes and Montgomery [2007], p. 233). And, as revealed somewhat in this section, there is a corresponding major difficulty of interpreting our interpretations in stochastics.

[29] Diaconis, Holmes and Montgomery [2007], p. 211.

[30] Diaconis, Holmes and Montgomery [2007], p. 211.

[31] Diaconis, Holmes and Montgomery [2007], p. 233.

[32] See notes 24 and 25.

frequencies of events or to propensities in the world. We can then use descriptions of truth-conditions to form simple slogans such as 'probabilities are frequencies.'[33]

I ask, here, that the reader hold off from philosophical debate. I am, instead, inviting the reader to attempt something that is not yet standard in philosophical reflection. I am asking that you attend to what *you* do. In that mode, we can ask the following question: Are the numerical probabilities obtained by Diaconis *et al.*[34] in some sense 'mind-independent'? I can imagine that a reaction of the authors and reviewers of the *SIAM* paper might very well be 'No.' As the reader will see below, that is also my answer. And, yet, the objectivist view asserts that there is something about their results that is mind-independent. So, let's explore this somewhat, by looking to results in the paper in the *SIAM* paper.

What results obtained by Diaconis *et al.* might be mind-independent? Is it description of Heads and Tails of coins that is mind-independent? However, obviously our mind is involved when we refer to images that we hold in sight and touch and use the names Heads and Tails. Perhaps it is counting that is mind-independent? But, certainly, when counting we are thinking. The number 0.51 is a fraction represented in the decimal system, going back to advances in understanding in ancient mathematics, and for us, also, not mind-independent. What of the subtle thinking when one asserts that

the chance of coming up as started is about .51[35]?

In a similar way, does one not use one's mind when one asserts that

(t)he classical assumptions of independence with probability 1/2 are pretty solid.[36]

Of course, understanding any part of these assertions involves one's mind.

You may suggest that inviting you to advert to elements of your own experience in these particular cases is not the meaning of 'mind-independent' in philosophy of physics. Some may suggest that interpreting

---

[33] See note 24.
[34] Diaconis, Holmes and Montgomery [2007].
[35] Diaconis, Holmes and Montgomery [2007], p. 211.
[36] Diaconis, Holmes and Montgomery [2007], p. 233.

0.51 needs to include that 0.51 is not just obtained through mental arithmetic, but that the number comes from experimental results, and that it reflects something about reality. To be sure, yes. But, how is that aspect of reality known? Is it not known by discovering the numerical probability 0.51? And, what are we to mean by that aspect of reality, if not that aspect of reality that we are talking about, namely, a numerical probability 0.51 discovered in doing this physics? I am only raising questions here. I am not attempting to say what a numerical probability of 0.51 means about reality. That is a difficult question that would need us to climb well beyond the present context[37]. I am, though, inviting your attention to a basic question: Is one to assert an interpretation of 'probability in physics' that does not bear out in one's experience in doing physics, when, say, counting Heads and Tails in a coin tossing experiment?

As mentioned, my main interest in this section is methodological. And for that, we have enough evidence for at least one feature of the problem to come somewhat into view, or rather, into mind: The *objectivist* view that asserts that probability is, in some, sense mind-independent[38], self-evidently does not bear out with what we do when we investigate relative actual frequencies and discover a numerical probability 0.51; or, in a similar way, are able to assert that, for natural flips, the numerical probability 1/2 is pretty solid[39].

---

[37] Among other things, one needs to identify key insights by which one grasps an 'ideal relative frequency,' also called 'numerical probability' (Lonergan [1992], Statistical Heuristic Structures, Sec. 2.4, pp. 76–91). See McShane [1970], for helpful discussion that reaches into the scientific and philosophic literature up to the late 20th century. McShane's book remains relevant: it indicates solutions to various problems; and many authors discussed there are still cited in the contemporary literature (e.g. Popper and Kuhn)). But, what is real? There is the problem of determining metaphysical equivalents (Lonergan [1992], Section 16.3.4). That will be a remote future achievement within generalized empirical method (ch. 6). One may reach a preliminary descriptive 'basic position' (Lonergan [1992], p. 413). But, potency, form and act are known through explanatory understanding. What will be needed will be a further major development, a basic position at the level of the times, a "come-about," described in *Insight*: "So it comes about ..." (Lonergan [1992], p. 537). See also ch. 6, note 43.

[38] The objectivist view is itself, of course, also not mind independent. This will need to be accounted for within a balanced view of probabilities. See note 37.

[39] Diaconis, Holmes and Montgomery [2007], p. 233.

However, the word 'probability' also is used in ways that are not numerical. So, let us now take a moment to consider a few aspects of the *subjectivist* view.

Subjectivist views, by contrast, take probabilistic statements to express degrees of belief. Thus, a fitting slogan is 'probabilities express credences.'[40]

A numerical probability is an answer to a question: '*What* is the *numerical* probability?'[41] The answer is a number (or an approximation to a number) between zero and unity. Is there also a question-form whose answer leads to *credence*? Perhaps?

One might ask, for example, 'Is the numerical probability 0.51 to be given credence?' In other words: '*Is it so* that 0.51 is the correct numerical probability?' Diaconis *et al.* appeal to mathematical and statistical analysis and experimental results to justify the claim. And, so as long as relative actual frequencies of samples used are found to not differ in a regular way from 0.51, then, 'Yes, 0.51 is probably correct.' In other words, 'Yes, 0.51 is to be given credence.'

Evidently, a numerical probability of 0.51 refers neither to a particular throw nor to any particular sample[42]. Nevertheless, we do sometimes need to think about particular throws and particular samples. As in the paper by Diaconis *et al.*, under appropriate circumstances, one might be able to forecast a single coin toss turning up Heads. Observe that, in that case, there is no fraction in a single throw. But, a question might be: Is the forecast to be given credence? Again, as above (when asserting credence of a numerical probability), notice, or rather self-notice that, here too, credence is reached by answering a question of the form *Is it so?*: '*Is it so* that a particular forecast is correct?' We also ask '*What is it* that is forecast?' The answer to that question is the event descriptively defined as Heads. But, '*Is it so* that the forecast of the event Heads is to be given *credence*?' In the special case described in the article by Diaconis *et al.*

---

[40] See note 24.
[41] See note 37.
[42] See note 37.

(where boundary conditions are known and the machine is well-engineered), the answer is 'Yes.' But, of course, even if the forecast is given credence, many things can happen that would lead to a different outcome, including vibrations in the machine caused by one of many earth tremors that tend occur in Stanford, California.[43]

Attempting to mix results helps further bring out that there are core differences between numerical probability and credence. How does one answer the question '*Is it so?*'? Is it with a numerical probability between zero and unity? Q: '*Is it so?*' A: '*0.51.*' Or, do we answer a question of the form '*What is the number* between zero and unity?' with a 'Yes,' 'No,' 'Probably Yes,' 'Probably No,' or 'Perhaps,' or other similar answers. Q: '*What is the number?*' A: '*Maybe.*'

Self-evidently, the combinations don't make sense. Note that this is not a logical argument, but an elementary and preliminary description: we raise and answer at least two different types of question, and get two different types of answer. In other words, based on our introductory exercises in self-attention, it is self-evident that we ask: 'What is it?' and 'Is it so?'[44] Our 'Is it so?' question presupposes our answer from the 'What is it?' question. But, when we answer the second question, we do not merely repeat the answer to the first question. Do we? Instead, we get to something new: a 'Yes,' or 'No,' or 'Probably Yes,' or 'Probably No,' or 'Maybe,' or 'I don't know.' And so we find that when the word '*probably*' is used in an answer to the second question, it is not a *number* in answer to a 'What is it?' question, but an *adjective* referring to one's confidence in one's 'Yes' or 'No.'

How does all of this play out in the paper by Diaconis *et al*? They discovered a fascinating[45] result: For what they defined to be natural flips, 0.51 is a good approximation to the *numerical probability* for a coin landing as it started. The number 0.51 is an answer to a 'What is it?'

---

[43] There were, for example, 399 earth tremors detected in Stanford, CA in 2013.

[44] To follow up on these preliminary and elementary observations, see, e.g., *Insight*, Lonergan [1992], ch. 10, Reflective Understanding. The book *Insight*, however, is an advanced text. For a pedagogical introduction, see Benton, Drage and McShane [2005], ch. 17, What-Questions; and ch. 18, Is-Questions; McShane [1973], ch. 3, The What-Question, ch. 4, The Is-Question. See also Figure E.1 in the Epilogue, and Lonergan [2001], Appendix A: Two Diagrams, pp. 319–323.

[45] Diaconies et al, p. 233.

question. However, is that numerical probability of 0.51 to be given credence? Here, we have an 'Is it so?' question. The numerical probability is, in fact, given credence by the authors. As mentioned above, appealing to a combination of mathematical and statistical analyses and experimental results, the answer given by Diaconis *et al* is: 'Probably yes!'[46], where the word *probably* here is an adjective expressing confidence in a result that was scrutinized by editors and at least six referees for *SIAM*.[47]

But, a common application of numerical probabilities is to particular cases. So, we need to ask what 0.51 might have to do with an individual toss in the case, say, where one does *not* have all of the boundary conditions. In such a case, how does the number 0.51 pertain to a single toss, when it is a natural flip and the coin starts out Heads? This, question, however, is better discussed in connection with subjectivist views.

---

[46] See note 44. For readers who intend to follow up on this initial foray: The method involves attending to one's acts and operations, and will lead to numerous clarifications. For instance, there is no appeal to 'hypothetical rational agents.' And, like any other numerical probability, discovering that unity is the numerical probability means that relative actual frequencies do not diverge from unity in a way that is too regular. [The meaning of 'not too regular,' i.e., randomness, becomes precise with inverse insight (Lonergan [1992], Section 4.2) contextualized within a measure space appropriate to the application.] In other words, even when numerical probability is unity, some relative actual frequencies can be, and are expected to be, less than unity. Indeed, if all relative actual frequencies were identically unity, if none were less than unity, statistical inquiry would not be needed and no numerical probability would be assigned. Similar comments apply when numerical probability is zero. That is, numerical probability being equal to zero is not equivalent to relative actual frequencies all being zero. Clarifications also will follow regarding philosophical interpretations of probability that define numerical probability as a 'degree of certainty.' We can already see from examples of this section that there is a fundamental error here. *Numerical probability* and *probability* (expressing certainty) are answers to different types of question, but with a homonym in common, namely, 'probability.' Numerical probability is an answer to a What-question, whereas 'degree of certainty' is an adjective, describing one's certainty in answering an Is-question. However, this is not to suggest that when numerical probabilities are known, there is no such thing as reasonable betting (McShane [1970]), the empirical significance of large numbers of samples or norms derived as mathematical limits. In fact, historically, a need to develop strategies for gambling was the catalyst for the beginnings of probability theory (see note 21).

[47] Diaconis et al. [2007], p. 234.

Recall that

> (s)ubjectivist views ... take probabilistic statements to express degrees of belief.
> Thus, a fitting slogan is 'probabilities express credences.'[48]

Now, our question, again, is does this view bear out in what we do in the work outlined Diaconis *et al.*? Does 'degree' in the expression 'degree of belief' refer, perhaps, to a numerical probability? But, numerical probability is an answer to a 'What-question.' Our confidence in being able to say, 'Yes, this numerical frequency is probably correct,' is supported by one's grasp of criteria met[49]. Or, as is normal in scientific collaboration, one may choose to not read the paper. For, of course, one cannot read and understand everything for oneself. In that case, one could, instead, choose to believe the authors, that their result 0.51 is correct.

But, the subjectivist view seems to be getting at something else. It associates a numerical probability with an act of belief. But, what is belief?[50] We may believe results of other authors. It is a choice, although usually not a blind choice, especially if one knows something about the authors and the journal. As we find in experience, then, belief is reached through a question of the form 'Will I believe?' And, the resolution of that question form is consent, or not, which, again, is not numerical.

Now, if the experimental arrangement were set up for a natural flip, and the coin started out as Heads, since the numerical probability for returning Heads is 0.51, it would be reasonable to bet on the outcome Heads[51]. But, observe that there is no relative actual frequency close to 0.51 in a single bet, neither in the number of events considered (one!), nor in the question 'Is it a reasonable bet?', nor in the further question, 'Will I make the bet?'[52] Moreover, it does not change either the credence of 0.51

---

[48] See note 24.

[49] See note 46.

[50] This needs continuation of work pointed to in note 44. See Lonergan [1992], Section 18.2. See also Benton, Drage and McShane [2005], ch. 20, What-to-do Questions and ch. 21, The Joy of Choice; and McShane [1973], ch. 6, Metaethics. In self-attention, we can describe four main question forms: What-is-it? Is-it-so? What-to-do? Is-to-do? See also Epilogue, Control of Meaning, Section E.2.

[51] See note 46. In particular, see McShane [1970].

[52] See also note 50.

or the reasonableness of the bet when Tails happens to be the outcome of a single toss, or even many subsequent tosses. Indeed, in this scenario, the authors of the article found that it takes large numbers of throws (of the order 250,000) for a bias to be observed in relative actual frequencies.

How, then, does the subjectivist view compare with our experience in working through results obtained by Diaconis *et al.*? Evidently, there is no experiential basis for calling a numerical probability a 'degree of belief.' If, however (as seems to be the usual application of the term) 'degree' refers to a statement such as, 'I am 51% confident,' then 'degree of 0.51' is only metaphor, serving as an adjective to express one's confidence, just like the words 'perhaps,' 'probably' and 'very probably.' And, one's degree of confidence in the bet does not change if, instead of Heads, a particular result is Tails.

I have only touched on a few details. The pointings to exercises needed in self-attention are elementary and preliminary, intended only as peeks and hintings at the possibility and relevance of such work[53]. And, we could keep teasing at the examples here, by bringing out further nuances when trying to see to what extent views (in the two main groupings mentioned above) match, or not, with our experience in getting the results of the paper by Diaconis *et al.* The reader, however, may not concur with my preliminary descriptions of, for example, four main question forms[54]. But, if you do not agree, is it perhaps because you have a better description, or assert the need for better description, of our question forms? What, then, are your descriptions of your question forms? Is there not a basic challenge here that needs to be met, namely, the challenge of asking about one's experience that is one's questioning?

In particular, does one's philosophical interpretation of probabilities in physics bear out in the subtleties of one's experience (which includes one's questions and insights) when obtaining and reaching probabilities in physics? If not, then, what does one mean by 'probabilities *in* physics'? However, while there are exceptions, the prevailing ethos in contemporary philosophy of physics does not require that one's interpretation of probabilities in physics be verifiable in one's questions, insights, and other experience in obtaining and reaching probabilities in physics. Instead, the

---

[53] See note 37.
[54] See note 50.

present tradition allows for, and promotes, ongoing debate about diverse interpretations of probabilities in physics, involving terms and criteria that do not need to bear out in one's experience when doing physics.

Inevitably, if one does not seek interpretation that bears out in one's experience in doing physics, one can be led to a notion of *interpreting a theory* where one's interpretation cannot be verified in anyone's experience in any world:

> What does it mean to *interpret*[55] a theory?... ; the interpreter will ask: "Under what conditions is this theory true?" and "What does it say the world is like?" The interpreter will then answer by specifying the class of worlds that make the theory true; or, a set of possible worlds according to the theory. Thus, interpretation, according to Gordon Belot, "consists of a set of stipulations which pick out a putative ontology for the possible worlds correctly described by the theory."[56]

But, as our preliminary exercises in self-attention reveal, in addition to wondering what might be so, yes, there *is* a natural follow-up question that can and does arise: *Is it so?*[57] Besides possibilities, what of actualities? In particular, what does physics in our world (as if that is not challenging enough!) mean by probabilities actually obtained or reached?

If one's effort to understand probabilities in physics is based in one's description of one's experience obtaining or reaching probabilities while doing physics, one will be taking steps toward a new balanced empirical method in science and philosophy of science[58]. However, like functional collaboration, emergence and normalization of the balanced empirical method belongs to a remote future. In the meantime, one can at least take a minimal position about one's experience. Are your efforts to interpret probabilities in physics partly based on your description of your experience when obtaining and reaching probabilities in physics? Or are your efforts to interpret probabilities in physics about something else?

You may recall that my main purpose here is not to debate the meaning of probability. Your interpretation of probabilities in physics may well be

---

[55] Italics in source text.

[56] Rickles [2008], pp. 7–8.

[57] See note 44. You may notice that it would be tricky indeed to explicitly hold to the negative view.

[58] See note 37.

different from mine. To the main point of this section, your position with regard to experience in physics also may differ from mine. But, whether expressed or not, evidently, we each have a position (or at least a prevailing orientation) with regard to experience in physics, within which we work out our interpretations of probabilities in physics.

In the hope of finding common ground, and of making progress together, might it not be helpful for us to make the effort to be as explicit as possible about our position with regard to experience in physics? Even if our positions differ, through making that effort, we would at least be getting to know more about sources of differences. But, there also will be at least some affinities, compatibilities and similarities. At the very least, we ask questions about interpretations of theories in physics, whatever that might mean. Evidently and (self-) evidently, then, there is a further task in evidence here. Whatever the methods and full reach of that further task eventually will be, it is something beyond what we have discussed so far in this book, for it is neither functional research, nor functional interpretation nor functional history.

4.3 Much More Is Needed: *Method* 250[59]

The previous section focused on the problem of interpreting probabilities in physics. But, of course, physics includes much more than probabilities. In contemporary High Energy Physics, probability distributions, events and occurrences all are provisionally defined in a Standard Model. There are also thermodynamics; biophysics; cosmology; technologies and engineering; education in physics; there is an emerging literature on physics, God and the sciences[60]; and so on. That there are many zones of inquiry in physics is well known. And, as the literature reveals, for each locus of inquiry X in physics, there can also be Philosophy of X[61]. Rickles describes something of the situation:

---

[59] See note 18.
[60] See, for example, Kragh [2004], Newsome [2011], Mann [2014] and Koperski [2015].
[61] Rickles [2008], p. 6.

There are many 'philosophy of ...' subjects, ... Indeed, most work in philosophy is of such a sort: the application to some specific field of enquiry or subject matter of the concepts, tools and methods of philosophy.[62]

But, this also means that the challenge of the previous section extends beyond the problem of interpreting probabilities in physics to the problem of interpreting all results in any locus X in physics, or in other sciences. As the reader may expect, my position minimally includes that my views about results in the sciences need to be verifiable in my experience obtaining results in the sciences. As in the previous section, then, an extended question for the reader is: Are your views about results in physics and the sciences verifiable in your experience obtaining results in physics and the sciences? Or, are your views about physics and the sciences about something else?

No doubt, there will be differences in our experience in physics and the sciences, in views, and in our positions. But, as mentioned in Section 4.2, there also will be at least some similarities and some affinities. For instance, if we speak about differences in our views, then at least we each have views of some kind. And that is something with which to begin. So, we might attempt to share our results; to speak about our respective positions; to the best of our ability work out and communicate our basic criteria; and attempt to do so in ways that might help each of us grow in our respective positions and eventually contribute to progress in the community. All of that would be very good. But, as history shows, collaboration is not easy. At this time, not only are there many views, but there are no standard methods or methodologies for resolving differences.

Bernard Lonergan, however, made remarkable progress in methodology. In what amounts to a one-page *précis* (*Method* 250), Lonergan mapped out the main features of what he called *functional dialectics*. However, that ultra-dense page is well beyond our present reach.[63]

---

[62] Rickles [2008], p. 7.

[63] McShane has written a book which, among other things, gives pointers on the meaning of one word of Lonergan's *Method* 250: 'comparison': McShane [2012].

To get a sense of the difficulty, imagine being a physicist in Galileo's time, trying to read a technically precise one-page abstract of the methods grounding the present Standard Model. Among other things, details would indicate layerings of 21st century cyclotron technologies and group theoretic gauge theory. But, the methods of Modern High Energy Physics have been centuries in the making. In a similar way, but with an even steeper learning curve, we can expect that functional dialectics (a fourth task in a future eightfold functional grouping of methodologically distinct tasks) also will be some time in the making.

In the meantime, though, we can extend our minimal position of the previous section, to a still minimal but somewhat more inclusive position, about experience in all of physics. In that extended minimal position, our growth in understanding what physics is needs to be verifiable in our experience doing physics.

For the purposes of this chapter, this provides us with additional (preliminary) data on the fourth functional task. For the purposes of the book, within that minimal position, we have evidence of commonalities running through all of the X's in physics, including the X that is called philosophy of physics. That is (as the first four chapters help reveal), by describing what we do in physics (including philosophy of physics) we find at least four fundamentally distinct tasks that, in this book, are called, respectively, functional research, functional interpretation, functional history and functional dialectics.

Are there other fundamentally distinct tasks? Obtaining evidence for four *forward* functional specialties will be the work of the next four chapters.

4.4 References

Beisbart, C. and Hartmann, S. (2011), *Probabilities in Physics* (Oxford University Press, Oxford).

Bell, E. T. (1937) *Men of Mathematics* (Simon and Schuster, New York).

Benton, J., Drage, A. and McShane, P. (2005) *Introducing Critical Thinking* (Axial Publishing, Vancouver).

Butterfield J. and Earman, H., eds. (2007) *Handbook of the Philosophy of Science, Philosophy of Physics* (Elsevier, Holland).

Cappelli, A., Castellani, E., Colomo, F. and Di Vecchia, P. (2012) *The Birth of String Theory* (Cambridge University Press, Cambridge).

Crull, E. (2013) Philosophy of Physics, *Analysis Reviews*, vol. 73, no. 4 October, pp. 771–784.

Diaconis, P., Holmes, S. and Montgomery, R. (2007) Dynamical Bias in the Coin Toss, *SIAM Rev.*, vol. 49, no. 2, pp. 211–235.

Frigg, R. (2008) A Field Guide to Recent Work on the Foundations of Statistical Mechanics, ch. 3 in Rickles, D. ed. (2008) *The Ashgate Companion to Contemporary Philosophy of Physics* (Ashgate Publishing Company. Burlington, VT), pp. 99–196.

Gross, D. (2005) The Future of Physics, *Internat. J. Mod. Phys. A*, vol. 20, no. 26, pp. 5897–5909.

Koperski, J. (2015) *The physics of theism: god, physics, and the philosophy of science* (Wiley, Hoboken).

Kragh, H. (2004) *Matter and Spirit in The Universe: Scientific and Religious Preludes to Modern Cosmology* (Imperial College Press, London) 2004.

Kuhlmann, M. and Pietsch, W. (2012) What is and Why Do We Need Philosophy of Physics?, in Kuhlmann, M. and Pietsch, W. eds., *Philosophy of Physics*, vol. 43, issue 2 of *J. Gen. Philos. Sci.*, pp. 207–375, pp. 209–214.

Lonergan, B. (1975) *Method in Theology* (Darton, Longman & Todd, London).

Lonergan, B. (1992) *Insight: A Study of Human Understanding*, vol. 3, *Collected Works of Bernard Lonergan*, eds. Crowe, F. E. and Doran, R. M. (University of Toronto Press, Toronto).

Lonergan, B. (2001) *Phenomenology and Logic: The Boston Lectures on Mathematical Logic and Existentialism*, vol. 18 in McShane, P., ed., *Collected Works of Bernard Lonergan* (University of Toronto Press, Toronto).

Mann, R. B. (2014) Physics at the Theological Frontiers, *PSCF*, vol. 66., no. 1, March, pp. 2–12.

McShane, P. (1970) *Randomness, Statistics and Emergence* (Gill and Macmillan, University of California). Also available at: http://www.philipmcshane.org.

McShane, P. (1973) *Wealth of Self and Wealth of Nations* (Exposition Press, New York).

McShane, P. (2012) *Method in Theology 101 AD 9011: The Road to Religious Reality* (Axial Publishing, Vancouver).

Rickles, D. ed. (2008) *The Ashgate Companion to Contemporary Philosophy of Physics* (Ashgate Publishing Company, Burlington, VT).

Newsome, W. T. (2011) Life of science, life of faith, ch. 36 in Chiao, R. Y. Cohen, M. L., Leggett, A. J., Phillips, W. D. and Harper Jr., C. L. eds., *Visions of Discovery: New Light on Physics, Cosmology, and Consciousness* (Cambridge University Press, Cambridge), pp. 730–750.

Rickles, D. (2014) *A Brief History of String Theory. From Dual Models to M-Theory* (Springer, Heidelberg).

Schwarz, J. H. (2012) The Early History of String Theory and Supersymmetry, https://arxiv.org/pdf/1201.0981v1.pdf, Jan. 4.

Smolin, L. (2006) *The Trouble with Physics, The Rise of String Theory, the Fall of a Science, and What Comes Next* (Houghton Mifflin, Boston).

Tannery, P. and Henry, C. (1904) *Oeuvres de Fermat*, vol. 2. (Gauthier-Villars et Fils, Paris).

Uffink, J. (2011) Subjective Probability and Statistical Physics, ch. 2 in, Beisbart, C. and Hartmann, S. (2011), *Probabilities in Physics* (Oxford University Press, Oxford), pp. 25–50.

Chapter 5

Functional Foundations

5.1 Introduction

The first four functional specialties will be 'past-oriented.' Sometimes types of events and occurrences are, in a sense, repeatable as, for example, in experimental physics. Or, again, anomalous text might be re-presented within a context. Thanks to the efforts of functional research, something that may be significant for progress is brought to the attention of functional interpreters. Functional historians will build on the results of functional interpreters. And, functional dialectics will strive to reach a luminous maximal progress-oriented embrace of all that has been and is – in the academic community, but inclusive of all that is known and is being done.

Some work, however, is 'future oriented.' Scholars work out truths and values by which to live and by which to guide future projects. There can be developments in understanding what further developments might be possible in given situations. Efforts can be made to sort out options and to discover how to bring ideas for development to diverse groups in science, the Academy, societies and the world. Prior to these tasks, however, there will be *functional foundations*, the first task in which the Academy properly turns toward the future.

Before going any further, a few comments here will, I hope, help avoid a possible misunderstanding. In the present context, the word 'foundations' is not used in the more familiar sense. In the more familiar sense of the word, there have been efforts to work out "foundations of quantum mechanics," "foundations of quantum field theory," and so on.[1]

---

[1] The tradition includes: David Hilbert's 6th Problem (1900), which calls for investigation into the possibility of mathematical axiomatization of physics. Well known books in the

There are also axiomatic foundations in philosophy, logic and the philosophy of science. In these contexts, the word 'foundations' is used when speaking of axioms, assumptions, postulates, and other logically first propositions.[2]

In the present context, *foundations* are, instead, 'fundamental directions.' The chapter consists, then, of descriptive searchings about fundamental directions, and about the task of making progress in fundamental directions[3].

Section 5.2 includes quotations from both Feynman and Ramanujan, indirect evidence of *foundations* in the sense intended in this chapter. Section 5.3 asks about *our* foundations. Section 5.3 includes a few personal reminiscences about my foundational background. I invite readers to attempt something similar for themselves. Even though indirect and descriptive, taken together, sections 5.2 and 5.3 help bring out that foundations are real. Section 5.4 is about Smolin's book, *The Trouble with Physics*[4]. Section 5.4 draws attention the fact that, among other things, the last chapters of Smolin's book invite foundational development in the sense described here. In other words, there are elements of Smolin's book that reveal the need and presence of a fifth task. However, that invitation in Smolin's book is not adverted to and mainly is only implicit. Section 5.5 is a sampling of foundational shifts that I envision for the Academy.

---

tradition include: axiomatic elements of Paul Dirac's book, first published in in 1930, *The Principles of Quantum Mechanics*, Dirac [1999]; John von Neumann's *The Mathematical Foundations of Quantum Mechanics*, von Neumann [1955]; Mackey's *The Mathematical Foundations of Quantum Mechanics*, Mackey [1963]; Part 1 of the book by Jauch [1968] is on axiomatization of mathematics needed, "Mathematical Foundations," Part 2, "Physical Foundations," is quantum mechanics in axiomatic form, while Part 3 is similar, but entitled "Elementary Particles."

[2] Lonergan [1975], pp. 269–270.

[3] For pointings regarding the complexity of the task in its maturity, see, e.g., McShane [2006], pp. 2–3. See also notes 23 and 26.

[4] Smolin [2006].

5.2 Intimations: Fundamental Directions in History

I begin with Richard Feynman (1918-1988). In memoirs, he wrote about his youth and his efforts to understand things.

> I learned very early the difference between knowing the name of something and knowing something.[5]

Note, also, that Feynman's later fundamental directions certainly were not the "cold scientific" interest represented in science-fiction and science-parody:

> I have a friend who's an artist and has sometimes taken a view which I don't agree with very well. He'll hold up a flower and say "look how beautiful it is," and I'll agree. ... At the same time, I see much more about the flower than he sees. I could imagine the cells in there, the complicated actions inside, which also have a beauty. I mean it's not just beauty at this dimension, at one centimeter; there's also beauty at smaller dimensions, the inner structure, also the processes. The fact that the colors in the flower evolved in order to attract insects to pollinate it is interesting; it means that insects can see the color. It adds a question: does this aesthetic sense also exist in the lower forms? Why is it aesthetic? All kinds of interesting questions which the science knowledge only adds to the excitement, the mystery and the awe of a flower. It only adds. I don't understand how it subtracts.[6]

I also think of *The Man who Knew Infinity*[7], Srinivasa Ramanujan (1887-1920). He too, early in life, found fundamental directions, in which he grew astonishingly, even though his life was relatively short[8].

> I have passed the Matriculation Examination and studied up to the First Arts but was prevented from pursuing my studies further owing to several untoward circumstances. I have, however, been devoting all my time to Mathematics and developing the subject.[9]

---

[5] This is a well-known quotation attributed to Feynman. See, e.g., http://www.richardfeynman.com/about/quote.html.

[6] BBC Interview, 1981: http://www.bbc.co.uk/programmes/p018dvyg/clips.

[7] Biographical film on S. Ramanujan, based on the book by Robert Kanigel of the same name, film written and directed by Matthew Brown, https://en.wikipedia.org/wiki/The_Man_Who_Knew_Infinity_(film).

[8] His story is well known. He died of tuberculosis contracted during his time in England.

[9] Berndt and Rankin [1995].

Something he said to a friend suggests that fundamental directions
of the adult Ramanujan were subtle indeed:

Sir, an equation has no meaning for me unless it expresses a thought of god.[10]

As the quotations point to, there are many fundamental directions,
including devotion to mathematics; physics; giving oneself to a life of
working out the marvels of chemistry; falling for the dynamics and beauty
of music and dance; an orientation to an unseen god; and aesthetics all
through. Both Feynman and Ramanujan seem to have been multi-
foundational persons with, though, at least one thing in common: both had
an openness and sustained devotion to understanding.

As the reader may gather, we all have our foundations, our fundamental
directions, our actual openness (or not), our attentiveness (or not), our
tendencies (or not) to ask and follow through with various types of
question and endeavor (or not).

5.3 My Foundations and Your Foundations

In the previous section, I was remembering Ramanujan and Feynman.
What, though, are *our* foundations?

For my part, I remember, as a teenager, learning High School physics
in Toronto. Mr. Gonzalez was an inspiring and gifted High School physics
teacher. He took us well into calculus and Newtonian physics. After that,
I was *into* mathematics and physics. A couple of years later, I encountered
a terribly short book by Michael Spivak, *Calculus on Manifolds*.[11] In a
mere 137 pages, the book subsumes centuries of mathematical
development, culminating in the generalized Stokes' theorem for k-forms
on a smooth manifold with boundary.[12]

---

[10] Famous statement to a friend, quoted in Ranganathan [1967], p. 88. Also quoted in
Pickover [2005], p. 1, but with no primary source indicated.
[11] Spivak [1965].
[12] Special cases include Green's theorem, Stokes' theorem and the divergence theorem, all
from classical physics.

I have learned much from many excellent teachers and books in mathematics and physics, and in other areas besides[13]. I single out these two experiences because the material will be familiar to readers in modern physics; and because my main point here is that both experiences – the two years learning High School physics with Mr. Gonzalez; and then later, struggling to read *Calculus on Manifolds* for the first time – each in different ways, refined, deepened and expanded my fundamental directions.

Several years later, during a post-doctoral year (1992-93) in Dublin, I was very glad to be admitted to a two-term seminar on group theoretic techniques in field theory, given by Lochlainn O'Raifeartaigh at the Dublin Institute for Advanced Studies. My Ph.D. was in C*-algebras, an area of mathematics that historically emerged from quantum mechanics. As I soon found out, however, because of gaps in my background in 20th century particle physics, O'Raifeartaigh's lectures were soon over my head. I worked hard, but wasn't able to catch up in two terms. Still, thanks to my effort to climb to the level of the seminar, my horizon[14] in modern physics expanded and was modernized. Of course, I also need to mention the writings of Bernard Lonergan and Philip McShane[15]. Throughout my career, ongoing efforts to digest their works has been resulting in ongoing transitions in my foundations, and indeed, in my horizon.

For the reader, you may also recall, perhaps, beginnings or changes in fundamental directions, or perhaps, epiphany days that changed the course of your development? You may also see now that this last paragraph of Section 5.3 goes with the final paragraph of Section 5.2. For, not only do we all have foundations, but, evidently, our foundations emerge, shift and grow. To introduce a name, let's call such changes *displacements*[16]. I have

---

[13] See, for example, notes 15 and 31.

[14] What is *horizon*? "(W)hat lies beyond one's horizon is simply outside the range of one's knowledge and interests: one neither knows nor cares. But what lies within one's horizon is in some measure, great or small, an object of interest and of knowledge" (Lonergan [1975], p. 236). See also note 31 in Chapter 2.

[15] Philip McShane, http://www.philipmcshane.org/.

[16] "Differences in horizon may be complementary, or genetic, or dialectical" (Lonergan [1975], p. 236). Taking McShane's lead here, I "prefer to use the word *displacement* which covers a (broad) class of events, including reversion, perversion, etc." (McShane [2002], p. 108).

found that sometimes displacements are gradual, through series of small shifts, over time. But, I also have found that there can be moments where, "suddenly as it were" (although, often following some prolonged effort), one may embrace some change in fundamental directions.

5.4 Discerning a Fifth Task

No doubt, foundations vary greatly, shaping ongoing progress and decline. At this stage in history, however, "few scientists think about foundational problems, and even fewer have ideas about them."[17] There are some, however, whose efforts implicitly reveal an effort to develop foundations[18].

This can be discerned, for example, in the last few chapters of Smolin's book[19], especially in the 20th chapter called What We Can Do For Science. Part of that 20th chapter is auto-biographical, telling something of Smolin's growth as a scholar, over decades.

He talks about how in the earlier part of his education, he learned science at a time when physics

> had been progressing so fast for so long that it was often taken as the model for how other kinds of science should be done.[20]

But, as Part IV of *The Trouble with Physics* shows, Smolin's questions, strategies and, generally, his context of concern, grew to include not only issues of contemporary theoretical physics, but physics of the past, present and future, areas and applications related to physics, and, indeed, peoples and societies of the world. In those last five chapters, he invites readers to a new holistic practical foundations for physics and all of science; for progress toward consensus about the need and possibility of experimental

---

[17] Smolin [2006], p. 329. Smolin does not speak of *foundations* in the sense intended in this chapter. His words here, however, fit and, as discussed below, the last part of his book implicitly invites *foundational* development.

[18] Again, I am not speaking here of logical foundations, but foundations in the sense of fundamental directions.

[19] Smolin [2006].

[20] Smolin [2006], p. xiii.

verification in physics; for progress in education, economics, university politics, ethics, cultures and history[21].

Smolin's invitation to an enlarged foundations is implicit. His explicit focus is, instead, mainly on consequences and applications. Also, various kinds of advice are offered:

> So I have some final words for different audiences. ... educated public, ..., those who make decisions about what science gets done, ..., (and) my fellow theoretical physicists.[22]

While there is an abundance of data revealing that Smolin grew in his foundations, we do not find passages in the book that speak directly about experiences that resulted in that growth, his rationale for embracing more inclusive foundations. There also is little discussion about how his new foundations emerge from and relate to his prior foundations, or discussion to promote the emergence of the fuller foundations in readers. There are, though, at least two features of the last chapters of Smolin's book that pertain to the present chapter: Implicitly, he calls for development in foundations (a) where one is to embrace best-to-date foundations already successful in the field; and also (b) where one is to hold (a) within a new more holistic heuristics for all of physics and endeavors related to physics.

Do we not then have data on a fifth task, a task that is part of looking toward the future? My purpose here is not to assess Smolin's views. My present focus continues to be on method. And, in the last chapters of Smolin's book, we find elements of methodological significance, namely, evidence of a fifth task. In its maturity, that fifth task will be *functional foundations*. Whatever that task eventually will look like, we already can see that it will include a filled out version of the (a) and (b) implicit in Smolin's work. That is, the fifth task will include: (a) taking up best-to-date foundations so far[23]; and (b) envisioning and embracing improvements or shifts of some kind.

---

[21] For present purposes, I leave it to the reader to go through those last chapters of Smolin's book. Eventually, though, functional research and functional interpretation would be needed.

[22] Smolin [2006], pp. 353–354.

[23] I am touching on two aspects of functional collaboration. (1) Once operative, part of the task of functional dialectics will be to communicate results forward to functional

The fifth task, then, will be creative, it will be for a better future, it will be *fantasy* about ways and directions not yet operative in the community. Of course, by saying 'fantasy' in this context, I am not referring to children's tales, but a task essential to progress:

> Without phantasy, all philosophical knowledge remains in the grip of the present or the past and severed from the future, which is the only link between philosophy and the real history of mankind.[24]

I am referring, then, to a serious and seriously demanding task of envisioning new or improved fundamental directions that, while not yet operative in the community, are concretely possible. It will be 'fantasy,' where the etymology[25] of the word includes the Greek word *phainein*, "to show, to bring to light."

Progress in foundations will have no precedent within the *acquis*. Such reaching will require relatively few citations. It will be a scientific and philosophic version of '*I* envision and embrace these new or improved directions.' The reaching and displacement beyond established ways will be fundamentally private. But, in order for one's foundational growth to benefit the community, there will need to be an effort to communicate something of one's new fundamental directions.[26]

---

foundations, a communication C_{56}. That is, such communication will be just one of 64 functionally distinct modes of internal communication C_{ij}, $i, j = 1, \ldots, 8$, that will be normative and normal in functional collaboration. See Section 8.5. (2) Best-to-date will include a control of meaning at the level of the times, in physics, biochemistry and biophysics, and so on. In practice, there will be *internal* communications between groupings of peers in the Tower of Able. A counting of functionally distinct modes of *internal* communication that accounts for this symmetry number is $8 + 7 + \ldots + 1 = [\frac{8 \times 9}{2}] = 36$.

[24] Marcuse [2009], p. 114.

[25] Etymology available in standard dictionaries. See, e.g., http://www.etymonline.com/. From Greek, *phantazesthai*, "picture to oneself"; *phantos* "visible," in late Greek "to imagine, have visions"; related to *phaos, phos* "light."

[26] In particular, there will be communications C_{56}. See note 23.

5.5 Something of What *I* Envision: Eightfold Experience and Eightfold Progress

I will not attempt to provide a full or detailed account of foundational progress that I envision for the Academy.[27] Instead, I will touch superficially on two aspects, in a way that is intended to be reachable from within the context of this introductory book.

Of course, a big part of what I envision is that physics and the entire Academy eventually embrace and grow in functional collaboration. This book is written as a small contribution to help toward that possibility. At this stage of the book, there are now five chapters on five distinct tasks. The next few chapters are intended to help readers round out to a preliminary awareness of (the pre-emergence of) all eight tasks; and eventually to consent to contributing to the emergence of functional collaboration.

How, though, are we to implement the new more effective collaboration? We will find out as we go. The new methodology will be an ongoing work-in-progress. Part of what is needed, however, already is in evidence. In both discernment and implementation, there is an unavoidable need of self-attention. This need has been increasingly evident in history and, in Chapter 4, was raised to (minimal) positional stance: adequate empirical method admits not only data of sense, but all experience in doing science.

Part of what I envision for the Academic community, however, is for us to eventually go much further, to go beyond minimal positions, to make progress in a balanced empirical method in all areas of scientific inquiry.

---

[27] That fuller account would include, for instance, heuristics of things with layerings of properties that are physical; chemical; and so on. Eventually, such heuristics will be needed in ongoing development in explanatory differentiation of foundations and horizons. See Lonergan [1975], Foundations, ch. 11. In the present-day Academy, an adequate heuristics of layerings has not yet been obtained. For Lonergan's solution, given in doctrinal brevity, see Lonergan [1992], ch. 8, Things, but then also the major transposition of that chapter to ch. 15, Elements of Metaphysics. See, in particular, Section 15.3, Explanatory Genera and Species. McShane has helpful work on the problem in McShane [1976], and several other later works. I explore details and a few examples from contemporary biochemistry and philosophy of science in *Invitation to Generalized Empirical Method*, Quinn [2017].

At first, results will be mainly descriptive and lacking in precision. Eventually, however, there will be what Lonergan called *generalized empirical method*.[28] Within generalized empirical method, there will be a control of meaning in which scholars will be, among other things, luminous about layerings of description and explanation, as well as to corresponding differences in what is intended, in being[29]. In particular, the future scholar will be (self-) luminous to the fact that (taking Feynman's words for my own),

> science knowledge only adds to the excitement, the mystery and the awe of a flower. It only adds.[30]

Finally, a few points of clarification about my foundational message. It is true that neither functional collaboration nor generalized empirical method are operative at this time. In that sense, what I envision and suggest for the Academy will be major shifts in history. But, at the same time, what I envision is not original.

So, I find it convenient, here, to appeal to McShane's distinction between Creative Learning and Creative Progress:

> the distinction between the creativity of learning and the creativity that pushes for progress[31].

Functional foundations will involve creative progress of front-line foundations persons who will help the community reach new positive displacements. However, what *I* presently envision is not so much the fruit of Creative Progress as the result of some years of sustained effort at Creative Learning, studying the works of two front-liners, Bernard Lonergan and Philip McShane[32]. Based on my experience so far, part of what I hope for the Academy will need to include a period of working together in Creative Learning. Generalized empirical method and functional collaboration will be fundamental transitions to a new

---

[28] See ch. 6.
[29] I am referring, here, to an advanced basic position. See note 36 of ch. 4.
[30] See note 6.
[31] McShane [2007], p. 66.
[32] See note 15.

effectiveness in human history. To give it a name[33], it will be a transition to a *third stage of meaning*.[34]

5.6 References

Berndt, B. C. and Rankin, R. A. (1995) *Ramanujan: Letters and commentary* (American Mathematical Society, Providence).

Dirac, P. A. M. (1999) *The Principles of Quantum Mechanics*, 4th ed. (Oxford University Press, Oxford).

Jauch, J.M. (1968) *The Foundations of Quantum Mechanics* (Addison-Wesley, Boston).

Lonergan, B. (1975) *Method in Theology* (Darton, Longman and Todd, London).

Lonergan, B. (1992) *Insight: A Study of Human Understanding*, vol. 3 in *The Collected Works of Bernard Lonergan*, eds. Crowe, F. E. and Doran, R. M. (University of Toronto Press, Toronto).

Mackey, G. W. (1963) *The Mathematical Foundations of Quantum Mechanics* (W. A. Benjamin, New York).

Marcuse, H. (2009) *Negations: Essays in Critical Thinking*, tr. from German by Shapiro, J. L. (Mayflybooks, London). First pub.: (1968), (Allen Lane, Penguin Press, UK).

McShane, P. (1976) *The Shaping of the Foundations: Being at home in the Transcendental Method* (University Press of America, Washington, D.C.).

McShane, P. (2002) *Pastkeynes, Pastmodern Economics. A Fresh Pragmatism* (Axial Publishing, Vancouver).

McShane, P. (2006) Ormerod's Dated Ecclesiology, *Joistings 19*, http://www.philipmcshane.org/.

McShane, P. (2007) *Method in Theology: Revisions and Implementations*, http://www.philipmcshane.org/.

Neumann, J. von. (1955) *The Mathematical Foundations of Quantum Mechanics* (Princeton University Press, Princeton).

Pickover, C. A. (2005) *A Passion for Mathematics: Numbers, Puzzles, Madness, Religion, and the Quest for Reality*, 1st ed. (John Wiley and Sons, Hoboken, NJ).

Quinn, T. (2017) *Invitation to Generalized Empirical Method* (World Scientific Publishing, Singapore).

Ranganathan, S. R. (1967) *Ramanujan, the Man and the Mathematician* (Asia Publishing House, Bombay).

Smolin, L. (2006) *The Trouble with Physics. The Rise of String Theory, the Fall of a Science and What Comes Next* (Houghton-Mifflin, Boston).

Spivak, M. (1965) *Calculus on Manifolds. A Modern Approach to Classical Theorems of Advanced Calculus* (Westview Press, Boulder, CO).

---

[33] See note 5.

[34] Lonergan [1975], Section 3.10, Stages of Meaning, pp. 85–99. Also, see Preface.

Chapter 6

Functional Doctrines and Policies

6.1 Introduction

Physics has come a long way in the last four centuries. What, though, of future progress? Readers may rightly point out that different areas in physics rely on different methods and have different goals. In one part of contemporary physics, the community searches for gauge theories verifiable in experimental data. Others, however, are differently concerned, seeking mathematical solutions, even when data is not available. In contemporary philosophy of physics, there is little sign of a coming convergence of views. There are, though, methods for which at least tacit agreement is given.

The point of this chapter, then, is not to compare different theories, philosophies and methods. It is, instead, to make beginnings in noticing certain commonalities, when physics moves toward new results. Are there any such commonalities at all in physics? Is there not at least one?: There is the judgment that *progress in physics is worthwhile.*

Still, just as areas in physics are diverse, so also are judgments about what is worthwhile, about which methods to employ, about truths by which to move toward future progress. Moreover, not only are such commitments diverse and often highly technical, but they have also shifted ground following major discoveries.

Sections 6.2–6.5 mainly are to provide preliminary data on judgments grounding future progress. These four sections regard, respectively, four broad subdivisions of the physics literature[1]. In Section 6.2, I look to

---

[1] As functional collaboration emerges, these groupings will shift and spread out within a developing functional control of meaning.

comments of some influential physicists. Section 6.3 regards physical theories. Section 6.4 is on physics and other sciences. Section 6.5 is on philosophy of physics. Section 6.6 makes a beginning toward envisioning a sixth functional specialty, *functional doctrines*. In Section 6.7, I give three doctrines that will help ground future progress in physics and other sciences.

6.2 Impromptu Reflections on Mathematics and Physics

Everyday description of lengths and times is, of course, not very much like mathematical understanding of such in geometry. What are the differences? Are there similarities? Thought on such questions goes back to ancient times and can be found, for example, in the works of Plato and Aristotle[2]. Galileo's 17th century discovery of the law of falling bodies, however, brought new focus to inquiry, in the context of an emerging physics.

Discovering and verifying a quadratic relation of measured lengths and times is, of course, remarkably different from observing an object in free-fall. Galileo's solution to this puzzle was to suggest the existence of *primary qualities* and *secondary qualities*. In modern rendering:

> Whereas *primary qualities* – such as figure, quantity, and motion – are genuine properties of things and are knowable by mathematics, *secondary qualities* – such as colour, odour, taste, and sound – exist only in human consciousness and are not part of the objects to which they are normally attributed.[3]

There is, in fact, ongoing philosophical debate about "primary and secondary qualities."[4] Let us look, instead, to something that happened in the physics community, following Galileo's discovery. A method emerged

---

[2] See, for example, Grattan-Guinness [1994].

[3] Epistemology: epistemology and modern science, *Encyclopedia Britannica*, https://www.britannica.com/. For Galileo's discussion of the matter, see, Galileo Galilei, *The Assayer, 1623*, in Drake [1957], pp. 274–277.

[4] Nolan [2011]. For a discussion of the error in the distinction, and of Lonergan's solution, see Quinn [in press], *Invitation to Generalized Empirical Method*, secs. 1.5–1.7.

and, implicitly at least, was sanctioned by a community of scholars[5]: *Progress in physics depends on finding mathematical correlations verifiable in measured lengths and times.* I am not suggesting that Galileo, or physicists after him, gave explicit expression to such a principle. It is, instead, to a judgment implicit in their work that I draw our attention.

That physics partly depends on mathematics and measurements is, of course, not a new observation[6]. Like Galileo in his time, many modern physicists (not only scholars in history or philosophy) also have pondered the matter, and related puzzles. In 1985, Feynman wrote:

> It is a small section only of natural phenomena that one gets from direct experience. It is only through refined measurements and careful experimentation that we have a wider vision. And then we see unexpected things: we see things that are far from what we would guess – far from what we could have imagined. Our imaginations stretched to the utmost, not, as in fiction, to imagine things which are not really there, but to comprehend those things which *are* there.[7]

More recently, Quigg wrote:

> Theoretical speculation and synthesis is valuable and necessary, but we cannot advance without new observations.[8]
>
> …
>
> (P)hysicists have known since the 1920's that to explain why a table is solid, or why a metal gleams, we must explore the atomic and molecular structure of matter. That realm is ruled not by customs of everyday life, but by the laws of quantum mechanics.[9]

I also need to draw attention to the writings of Penrose, a small sampling of which is as follows:

---

[5] "The design of experiments to discover new mathematical laws comes after Galileo's time" (Drake [1970], p. 44).

[6] In the history and philosophy of science, there are long-established literatures on the topic. See, e.g., Nolan [2011], and references therein.

[7] Feynman [1985], pp. 127–128. Italics in source text.

[8] Quigg [2006], p. 115.

[9] Quigg [2006], p. 116. There are Eddington's famous questions about "two tables," one "scientific" and one "everyday," in his 1927 Gifford Lectures at the University of Edinburgh. See Eddington [1929], Introduction, pp. ix–xvii.

We cannot get any deep understanding of the laws that govern the physical world without entering the world of mathematics.[10]

...

(A)esthetic criteria are fundamental to the development of mathematical ideas for their own sake, providing both the drive towards discovery and a powerful guide to the truth.[11]

...

(T)he criteria of science are not those of democratic governance. It is right and proper that minority activities should not suffer merely by virtue of the fact that they are in the minority. Mathematical coherence and agreement with observation are far more important. But, can we ignore the whims of fashion altogether?[12]

...

(T)here is little doubt in my own mind but that mathematical aesthetics must be an important driving force in addition to physical insight. ... In accordance with this, progress towards a deeper physical understanding, if it is not able to be guided in detail by experiment, must rely more and more heavily on an ability to appreciate the physical relevance and depth of mathematics, and to 'sniff out' the appropriate ideas by use of a profoundly sensitive mathematical appreciation.[13]

In the same publication year as the book by Penrose, Smolin claims the following:

As science is based on experiment, we cannot give a convincing answer to what we have learned, if we are not able to explain how our ideas and calculations will be tested.[14]

Yau and Nadis began their 2010 book with an apparently elementary observation:

---

[10] Penrose [2004], p. xix.
[11] Penrose [2004], p. 22.
[12] Penrose [2004], p. 1018.
[13] Penrose [2004], p. 1026.
[14] Smolin [2004], p. 494.

We're familiar with travel in three basic directions: north or south, east or west, and up or down. (Or, equivalently, left or right, backward or forward, up or down.)[15]

The book then rapidly lifts to a discussion of sophisticated geometries in contemporary cosmology. Regarding the emphasis of the book, Yau writes:

I'm not playing up the geometry angle because it happens to be the main thrust of this[16] book. It's also vital to the endeavor about which we speak. For one thing, we cannot describe forces – an essential part of the Standard Model (and of any purported theory of nature) – without geometry.[17]

As mentioned above, discussion of these issues is not new. What is relatively new here is an observation: In each of the quotations just given, there is a judgment expressed regarding what for each author would be *future* progress. Moreover, those judgments were part of their ongoing efforts to make new progress.

There are compatibilities in the examples just given. But, of course, not all judgments about future progress are compatible. In some cases, we find open conflict. A well-known case in point is the ongoing conflict about whether or not string theory is correct[18]; and, indeed, whether or not methods of string theory are, or are not, methods of empirical science. As Castelevecchi observes, there has been an ongoing general agreement (in fact, among a large majority of the physics community) that physical theories need to be in some way supported by experiment or observation. On the other hand, proponents of string theory seek to determine

internal consistency of a theory or the absence of credible alternatives, to update estimates, instead of basing those revisions on actual data.[19]

String theory, then, also goes forward on claims about method, to the effect that a physical theory does not necessarily need to be verifiable in experiment or observation.

---

[15] Yau and Nadis [2010], p. 3.

[16] Referring to their book, Yau and Nadis [2010].

[17] Yau and Nadis [2010], p. 202.

[18] Casetelvecchi [2015]. See also the description of the debate given by Gross, in ch. 4, note 8.

[19] Casetelvecchi [2015].

To continue gathering data on judgments regarding future progress, let's look again to the example mentioned above. More specific than claims about "theory and observation," there is an operative principle that started with Galileo: *Progress in physics depends on finding mathematical correlations verifiable in measured lengths and times.*[20] The principle is also evident in: Newton's work and his appeal to data on lunar and planetary orbits; the efforts of Ørsted, Ampère, Faraday, Ohm; the series of experiments and theoretical developments that led up to, and include, Maxwell's system of space-time partial-differential equations for an electromagnetic field; the context leading up to, and including special relativity, and eventually also general relativity; quantum mechanics; the rise of gauge theory following Weyl's 1929 paper[21]; and ongoing experimental and theoretical work contextualized by the present Standard Model.

In other words, while theories and contexts have changed, the developments mentioned have not been without relatively permanent operative premises about progress. Whatever historical or philosophical analysis eventually brings to light about methods and theories, and whatever the following words eventually may come to mean, implicitly at least, there have been recurring judgments regarding the need to search for *mathematical correlations verifiable in measured lengths and times.* Even if the principle has been provisional, even if manifesting differently at different times in different contexts, and even if (as some suggest), it is to be eventually replaced, so far, the commitment has been an operative premise in mainstream physics since Galileo's time.

I have been talking about community-wide commitments, operative and influential in ongoing progress in physics. Before going further, it will be convenient to introduce the name 'doctrine.' Contemporary standard English dictionary definitions of 'doctrine' include: "something that is taught"; "a principle or body of principles in a branch of knowledge";

---

[20] Exercises toward implementation of generalized empirical method (the second doctrine of Section 6.7) lead to more precise identification. When verified, one is grasping a "correlation of correlations of correlations" (Lonergan [1992], p. 271). See second last paragraph of Section 6.7; and Epilogue.

[21] Weyl [1929]. Weyl's 1929 article is reproduced in ch. 5, Weyl's Classic, O'Raifeartaigh [1997].

"a principle of law established through past decisions"; "a system of belief"; "a dogma"; "a policy"; "a set of strategies."[22] These usages are not meant to limit or prescribe doctrines in physics. The word 'doctrine,' though, provides a convenient name for relatively permanent descriptive truths.

6.3 Physical Theories

In the previous section, I pointed to a few of the well-known advances in physics. In fact,

(h)istorically, there have been several major watersheds for physics.[23]

Where Galileo investigated motions on Earth, Kepler investigated motions of planets in the solar system[24]. Newton went on to discover a universal theory of gravitation. Initially, there was the effort needed to learn what Newton meant. This included getting to know his *method of fluxions*, as well as his (inverse[25]) insight to focus not on velocity but on change in velocity. Within the Newtonian system, one could both explain free-fall and re-derive Kepler's laws of planetary motion. Follow-up developments were extensive and include, for example, classical fluid dynamics, continuum mechanics, and much of 20th century civil engineering.

Other watersheds include Maxwell's discoveries and, of course, special relativity. Both of these were subsumed by quantum electrodynamics. Quantum electrodynamics survives in the electroweak

---

[22] See, for example, http://www.merriam-webster.com/.

[23] Fraser [2006], pp. 1–2.

[24] Kepler's work also contributed to the emergence of the new method. Although, Kepler himself seems to have moved within a weave of old and new operative premises. Like Galileo, his results emerged through (an extraordinary decades-long effort) studying tables of (astronomical) measurements (given to him by his colleague Tycho Brahe). Kepler's second law of planetary motion, about area sweeps in time intervals, subtly hints at the possibility of a dynamics. But, his results also emphasize imaginable static conic sections. Future functional interpretation will be needed to interpret Kepler's results.

[25] Lonergan [1992], sec. 2.4, Inverse Insight, pp. 43–50.

theory. Weyl elevated gauge invariance to a fundamental principle[26]; and there emerged the now familiar general principle called *gauge theory*[27].

I am just skimming across recent history here. I am not trying to solve problems but am raising questions: What are doctrines of past and present physics? How are they related? In what ways do doctrines for quantum electrodynamics include or subsume earlier doctrines of classical electrodynamics? Were there minor or major adjustments or shifts in doctrines? Are there doctrines proper to theories, and proper to functional specialties? How do doctrines about progress, about methodology and methods relate to doctrines regarding particular theories, geometries and boundary conditions? And so on, and so forth.

These questions call for functional research, interpretation, history and dialectics. Note, in particular, that a history of doctrinal developments would have a related but different focus from a history that emphasizes, say, conceptual developments[28]. After all, doctrines and concepts are not identical classes of events in history. In light of Chapter 3, then, we can begin to envisage sequences of related systems of doctrines, their emergence and survival (or not), where survival of a doctrine means some kind of transposition to a new context.

In the meantime, perhaps enough has been said to bring out that, at least implicitly, there are future-oriented relatively permanent descriptive truths that ground ongoing progress in physics. Many questions arise. But, in accordance with the introductory nature of this book, the next section will keep to the work of gathering data.

6.4 Physical Sciences and Community Enterprise

The obvious fact is that

physics has ... become highly multidisciplinary[29].

---

[26] O'Raifeartaigh [1997], p. 6.

[27] O'Raifeartaigh [1997]; and O'Raifeartaigh, L. and Straumann, N. [2000]. See also, Taylor [2001].

[28] See, for example, Cao [1997].

[29] Fraser [2006], p. 7.

This already was discussed somewhat, in the Introduction to this book. In this chapter, however, the focus is on doctrinal presence. Consider, then, the following future-oriented statement:

> (I)t now seems clear that any systematic investigation of the sciences requires the tools of a wide range of disciplines, and that to understand this complex family of enterprises it is essential that the contributions of practicing scientists, of social scientists, and of science studies scholars from a range of disciplines, should inform one another.[30]

At this time, there are well more than a hundred recognized main focus-areas in physics, and the number is growing. The scientific tradition does not yet ask that scholars advert to doctrines. Nevertheless, even superficial review reveals that doctrinal premises ground ongoing work in the entire interdisciplinary physics community. This is pointed to in the quotation of note 30 of this chapter. Other examples are evident in cases where future-oriented doctrines become explicit, even if not adverted to as such.

For example, Roger Penrose writes:

> I believe that it is more important than ever, in today's technological culture, that scientific questions should not be divorced from their moral implications.[31]

There is Safinya's claim regarding future progress in medicine:

> The interdisciplinary nature of the field means that significant progress requires a thorough understanding of the relevant biological issues in addition to the day-by-day collaboration and exchange of ideas among physical scientists, biologists, and medical researchers.[32]

The conclusion in the following quotation reveals a doctrinal presence in socio-ecology:

> The most important first step is to agree that an ever-rising energy and material throughput is not a viable option on a planet that has a naturally limited capacity to absorb environmental by-products of this ratcheting process. To invert Lotka's dictum, we must so operate as to stabilize the total mass of the organic system, to limit the rate of circulation of matter through it, and to leave an unutilized residue of matter and available energy in order to ensure the integrity of the biosphere.[33]

---

[30] Nelson [2002], 326–327.
[31] Penrose [2004], p. 22.
[32] Safinya [2006], p. 442.
[33] Smil [2011], p. 728.

6.5 Philosophy of Physics

As already mentioned, in the philosophy of physics there has been little evidence of an emerging standard model. There are, though, clusterings of operative future-oriented premises, implicit in efforts to obtain new results within particular views. These views can be associated with names such as materialism, realism, convergent realism, scientific realism, anti-realism, idealism, conceptualism, constructivism, determinism, indeterminism, emergentism, and so on. And, new combinations and varieties of operative premises (some of which are mutually incompatible) continue to emerge[34].

If, though, we look to *methods* in contemporary philosophy of physics, some commonalities can be discerned. Tim Maudlin points to this, in his discussion of what (contemporary) philosophers of physics do:

> Philosophers strive for conceptual clarity. Their training instills certain habits of thought – sensitivity to ambiguity, precision of expression, attention to theoretical detail – that are essential for understanding what a mathematical formalism might suggest about the actual world. Philosophers also learn to spot the gaps and elisions in everyday arguments. [35]

Maudlin's description bears out in, for instance, Healey's book on the conceptual foundations of contemporary gauge theories[36]:

> What is needed is ... careful logical and philosophical examination of the conceptual structure and broader implications of contemporary theories.[37]
>
> ...
>
> (T)he key interpretative question addressed here is the following: *What beliefs about the world are (or would be) warranted by the empirical success of this (gauge) theory?*[38]

---

[34] See ch. 4.
[35] Maudlin [2015], par. 10.
[36] Healey [2007].
[37] Healey [2007], p. ix.
[38] Healey [2007], p. xv. Italics in source text.

Later in the book, there is a definition of *physical theory*:

A physical theory specifies a set of models – mathematical structures – that may be used to represent various different situations, actual as well as merely possible, and to make claims about them.[39]

These are, of course, just a few fragments from a vast literature. Still, they are representative of prevailing methods in contemporary philosophy of physics, that is, methods that focus on: formal and symbolic logic; terminologies; mathematics; conceptual analysis and commonsense arguments. In other words, the present philosophic tradition reveals an ongoing operative commitment to doctrines that *ongoing progress in philosophy of physics will be through speculative debate about images, concepts, terms and worlds that might or might not be.* Even though contemporary philosophical views are not yet approaching common doctrines about views, what is in evidence, then, at least implicitly, are commonalities in doctrines about method.

6.6   Hintings of *Functional Doctrines*

The first five sections of this chapter help bring out that there are diverse doctrines that ground ongoing development in physics. Three of the perhaps most elementary are: (1) Progress in physics is worthwhile; (2) Progress in physics is collaborative; and (3) Working in physics is worth a life. Also evident, however, is that, at any stage of development, doctrinal presence is subtle, complex, historical, and goes well beyond these three elementary doctrines.

What would be helpful would be to make progress in identifying and organizing main scientific-philosophic doctrines that are to ground future progress. As evident in the examples given in this chapter, such an effort will need to build on foundational work that identifies fundamental

---

[39] Healey, [2007], p. 150.

directions. But, that means that we have preliminary evidence of the need and possibility of a sixth task, in this book called *functional doctrines*.

A few aspects of this task can be anticipated. Past doctrines sometimes are retained or subsumed within later developments. The sixth task, then, will need to identify doctrines in developmentally related contexts.

Given the complexity and "layerdness"[40] of interdisciplinary physics and science, we can expect this to be intrinsic in up-to-date doctrines.

There would seem to be many kinds of doctrine. Doctrines can be about boundary conditions, experience, theories, methods and methodologies, and more. Eventually, there will be doctrines, too, about progress in functional research, functional interpretation, functional history, and so on. That is, there will be progress in doctrines about the entire functional collaboration.

The task called functional doctrines, however, will be its own main task, the purpose of which will include identifying and organizing not just two or three doctrines, or even any discrete collection of doctrines. Being up-to-date and oriented toward future progress, functional doctrines will take special advantage of input from functional foundations and functional dialectics. The sixth task, though, will not be looking to the past. With a control of meaning and best-to-date fundamental directions, the task called functional doctrines will (among other things) seek to identify sequences of variously overlapping "layerings" of relatively permanent descriptive pragmatic truths and values that are to ground future progress.

6.7 Three Doctrines

As readers might expect me to include, a doctrine I suggest is: *Functional specialization will be a normative eightfold functional division of labor and will lead to cumulative and progressive results*[41].

---

[40] I use the word "layered" loosely and descriptively. See ch. 9, on aggreformism. There are "layerings" of physics; chemistry; botany; zoology; and the human sciences. Control of meaning will require a development in empirical method. See note 43. See also the second doctrine of Section 6.7.

[41] Lonergan [1971], pp. 4, 5.

Toward a second doctrine, note first that making progress in (self-) detecting eight main tasks also means progress in self-attention. So, elementary progress in describing the eight main tasks also means elementary progress toward a *balanced empirical method* in the sciences. The possibility of such a method was first indicated by Bernard Lonergan, in his book *Insight*:

> But it may be urged that empirical method, at least in its essential features, should be applicable to the data of consciousness no less than the data of sense. Now on this matter a great deal might be said, ... We have followed the common view that empirical science is concerned with sensibly verifiable laws and expectations. If it is true that essentially the same method could be applied to the data of consciousness, then respect for ordinary usage would require that a method which only in its essentials is the same be named a generalized empirical method.[42]

Later, in 1974, Lonergan gave the balanced empirical method a more precise definition. Progress in working out the meaning of Lonergan's definition will be for future functional interpreters. For us, the definition is descriptive, relatively permanent, pragmatic, doctrinal for future progress in the Academy, and in particular, in physics:

> Generalized empirical method operates on a combination of both the data of sense and the data of consciousness: it does not treat of objects without taking into account the corresponding operations of the subject; and it does not treat of the subject's operations without taking into account the corresponding objects.[43]

---

[42] Lonergan [1992], 95–96. The 1992 edition is an edited version of the original 1957 publication. See 1992 edition for details on publication history.

[43] Lonergan [1985], p. 141. Eventually, it will be simply be adequate Empirical Method (Lawrence, [2007], p. 131). For inroads to the more advanced work, see McShane [1970], [1976] and [1980], and Quinn [in press], *Invitation to Generalized Empirical Method*. Scholars already familiar with Lonergan's work will recall the brief description of a *basic position* (Lonergan [1992], p. 413). In the sciences and philosophy of science, development in control of meaning will need to include progress toward identifying central and conjugate potencies, forms and acts and metaphysical equivalences. A *basic position at the level of the times* will involve a '*comeabout*,' as indicated in Lonergan [1992], p. 537. Present confusions that come from not distinguishing defining relations and boundary conditions will need to be gradually sorted out. Dense doctrinal pointers are in Lonergan [1992], Elements of Metaphysics, ch. 15, pp. 456–511.

A third doctrine regards one of the objectives of theoretical physics.[44]

> The abstract formulation, then, of the intelligibility immanent in Space and in Time is, generically, a set of invariants under transformations of reference frames, and specifically, the set verified by physicists in establishing the invariant formulation of their abstract principles and laws.[45]

Note finally that, in this chapter, I am not doing *functional doctrines*. In particular, I make no attempt to relate the three doctrines above. However, I expect that the importance of the second and third doctrines will become increasingly (self-) evident as the community makes progress toward a functional division of labor indicated in the first doctrine[46].

While there will be the work of sorting out best-to-date basic scientific-philosophic doctrines for future progress, other kinds of question arise. For example, with best-to-date doctrines in mind, what are possibilities and probabilities for future progress? Again, what are possibilities and probabilities for recovery from contemporary difficulties or confusions (where 'contemporary' is, of course, a relative term)? These and other questions lead to the next chapter on functional systematics.

6.8 References

Amaldi, U. (2006) *The New Physics for the Twenty-First Century*, ed. Fraser, G., Chapter 19, Physics and Society (Cambridge University Press, Cambridge), pp. 505–531.

Barrow, J., Davies, P.C.W., Harper, C. L. Jr. (2004) *Science and Ultimate Reality. Quantum Theory, Cosmology and Complexity* (Cambridge University Press, Cambridge).

Callender, C. and Hoefer, C. (2002) Philosophy of Space-Time Physics, ch. 9 in Machamer, P. and Silberstein, M., eds., *The Blackwell Guide to the Philosophy of Science*, (Blackwell Publishers, Massachusetts), pp. 173–198.

Casetelvecchi, D., (2015) Feuding physicists turn to philosophy for help, *Nat. Mag., Phys.*, vol. 528, issue 7583, December 23, pp. 446–447.

Coa, Tian Yu (1997) *Conceptual Developments of 20th Century Field Theories* (Cambridge University Press, Cambridge).

---

[44] Elsewhere (Quinn [2016]) I called this Theorem A of Lonergan's ch. 5 in *Insight* (Lonergan [1992]).

[45] Lonergan [1992], ch. 5, Space and Time, p. 174.

[46] See Epilogue.

Dirac, P. A. M. (1939) The relation between mathematics and physics, *Proc. R. Soc. Edin.*, 59, pp. 122–129.

DiSalle, R. (2006) *Understanding Space-Time. The Philosophical Development of Physics from Newton to Einstein* (Cambridge University Press, Cambridge, USA).

Drake, S. tr. (1957) *The Assayer* (1623), in *Discoveries and Opinions of Galileo* (Doubleday, New York).

Drake, S. (1970) *Galileo Studies* (The University of Michigan Press, Ann Arbor, MI).

Eddington, A. S. (1929) *The Nature of the Physical World* (Macmillan Co., New York). See also, https://archive.org/details/natureofphysical00eddi. The lectures also are available at http://www-groups.dcs.st-and.ac.uk/history/Extras/Eddington_Gifford. html.

Einstein, A. (1905) On the Electrodynamics of Moving Bodies, *Ann. Phys.* 17, pp. 891–921. Reproduced in The Library of Congress, and available online at http://einsteinpapers.press.princeton.edu/vol2-trans/154.

Feynman, R. (1985) *The Character of Physical Law*, 12th printing (M.I.T Press, Cambridge, Massachusetts).

Feynman, R. (1999) *The Pleasure of Finding Things Out* (Helix Books, Cambridge, Massachusetts).

Fraser, G., ed. (2006) *The New Physics for the Twenty-First Century* (Cambridge University Press, Cambridge). ·

Goddard, P. ed. (1998) *Paul Dirac: The Man and His Work* (Cambridge University Press, Cambridge).

Grattan-Guinness, I. (ed.), (1994) *Companion Encyclopedia of the History and Philosophy of Mathematical Sciences* (Johns Hopkins University Press, USA).

Healey, R. (2007) *Gauging What's Real, The Conceptual Foundations of Contemporary Gauge Theory* (Oxford University Press, Oxford).

Jeraj, R. (2009) Future of Physics in Medicine and Biology, *Acta Oncol.*, 48, pp. 178–184.

Lambert, P. and McShane, P. (2010) *Bernard Lonergan, His Life and Leading Ideas* (Axial Publishing, Vancouver).

Lawrence, L. (2007) The Ethics of Authenticity and the Human Good, in Liptay J. J. Jr. and Liptay, D. S. (2007), eds., *The Importance of Insight: Essays in Honor of Michael Vertin* (University of Toronto Press, Toronto), pp. 127–150.

Lonergan, B. (1971) *Method in Theology* (Darton, Longman and Todd, London).

Lonergan, B. (1985) *A Third Collection*, ed., Crowe, F. E. (S. J.) (Paulist Press, New York).

Lonergan, B. (1992) *Insight: A Study of Human Understanding*, vol. 3 in Crowe, F. E. and Doran, R. M. eds., *Collected Works of Bernard Lonergan* (University of Toronto Press, Toronto).

Machamer, P. (2002) A Brief Historical Introduction to the Philosophy of Science, ch. 1 in Machamer, P. and Silberstein M., eds. (2002), *The Blackwell Guide to the Philosophy of Science* (Blackwell, Massachusetts), pp. 1–17.

Mann, R. B. (2014) Physics at the Theological Frontiers, *PSCF*, vol. 66, no. 1, pp. 1–12.

Maudlin, T. (2015) *The Nature of Reality* (NOVA), Thought Experiments, Why Physics

Needs Philosophy, Thursday, April 23. http://www.pbs.org/wgbh/nova/blogs/physics/2015/04/physics-needs-philosophy/.

McShane, P. (1970) *Randomness, Statistics and Emergence* (University of Notre Dame Press, Notre Dame, IN).

McShane, P. (1976) *The Shaping of the Foundations: Being at home in the Transcendental Method* (University Press of America, Washington, D. C.).

McShane, P. (1980) *Lonergan's Challenge to the University and the Economy* (University Press of America, Washington, D. C.).

McShane, P. (2001) Elevating *Insight*: Space-Time as Paradigm Problem, *Method* vol. 19, no. 2, pp. 203–230.

McShane, P. (2007) The Importance of Rescuing *Insight* ch. 12 in Liptay, J.L. Jr. and Liptay, D.S., eds., *The Importance of Insight, Essays in Honour of Michael Vertin* (University of Toronto Press, Toronto), pp. 199–225.

Nelson, L.H. (2002) Feminist Philosophy of Science, ch. 15 in Machamer, P. and Silberstein, M., eds., *Blackwell Guide to the Philosophy of Science* (Blackwell, Malden, Massachusetts), pp. 312–331.

Newsome, W. T. (2011) Life of science, life of faith, ch. 36 in Chiao, R. Y. *et al*, eds., *Visions of Discovery. New Light on Physics, Cosmology, and Consciousness* (Cambridge University Press, Cambridge), pp. 730–750.

Nolan, L. ed. (2011), *Primary and Secondary Qualities: The Historical and Ongoing Debate* (Oxford University Press, Oxford).

O'Raifeartaigh, L. (1997) *The Dawning of Gauge Theory* (Cambridge University Press, Cambridge).

O'Raifeartaigh, L. and Straumann, N. (2000) Gauge Theory: Historical origins and some modern developments, *Rev. Mod. Phys.*, vol. 72, issue 1, Jan., pp. 1–23.

Penrose, Roger (2004) *The Road to Reality. A Complete Guide to the Laws of the Universe* (Alfred A. Knopf, New York).

Quigg, C. (2006) *The New Physics for the Twenty-First Century*, ed. Fraser, G., ch. 4, Particles and the Standard Model (Cambridge University Press, Cambridge), pp. 86–118.

Quinn, T. (2016) Interpreting Lonergan's 5th Chapter of *Insight*, ch. 2 in Brown, P. and Duffy, J. eds., *Seeding Global Collaboration* (Axial Publishing, Vancouver), pp. 29–44.

Quinn, T. (2017) *Invitation to Generalized Empirical Method* (World Scientific Publishing, Singapore).

Redhead, M. (1987) *Incompleteness, Nonlocality, and Realism: A Prolegomenon to the Philosophy of Quantum Mechanics* (Oxford University Press, Oxford).

Safinya, C.R. (2006) Biophysics and biomolecular materials, ch. 16 in, Fraser, G., ed., *The New Physics for the Twenty-First Century* (Cambridge University Press, Cambridge), pp. 405–443.

Silberstein, M. (2002) Reduction, Emergence and Explanation, ch. 5 in Machamer, P. and Silberstein, M., eds., *The Blackwell Guide to the Philosophy of Science* (Blackwell, Malden, Massachusetts).

Smil, V. (2011) Science, energy, ethics, and civilization, ch. 35 in Chiao, R. Y. *et al*, eds., *Visions of Discovery. New Light on Physics, Cosmology, and Consciousness* (Cambridge University Press, Cambridge), pp. 709–729.

Smolin, L. (2004) Quantum Theories of Gravity: results and prospects, ch. 22 in Barrow J.D., Davies, P. C. W. and Harper, C. L. H. Jr., eds., *Science and Ultimate Reality. Quantum Theory, Cosmology and Complexity* (Cambridge University Press, Cambridge), pp. 492–527.

Smolin, L. (2006) *The Trouble with Physics. The Rise of String Theory and the Fall of a Science, and What Comes Next* (Mariner, Houghton-Mifflin, Boston).

Suijlekom, W. D. van (2015) *Noncommutative geometry and particle physics* (Springer, Dordrecth).

Taylor, J.D. (2001) *Gauge Theories in the Twentieth Century* (Imperial College Press, London).

Weyl, H. (1929) Elektron und Gravitation, *Zeit. f. Phys.*, 56, pp. 330–352.

Weyl, H. (1997) Weyl's Classic, 1929, ch. 5 in O'Raifeartiagh, L., *The Dawning of Gauge Theory* (Cambridge University Press, Cambridge), pp. 107–144.

Yau, Shing-Tung and Nadis, S. (2010) *The Shape of Inner Space, String Theory and Geometry of the Universe's Hidden Dimensions* (Basic Books, NY).

Chapter 7

Functional Systematics: Systems-Planning

7.1 Introduction: Caring for Our PUP

I start this chapter with a question that, admittedly, may at first seem unusual for a book about physics. I ask that we think, concretively, of a sick greyhound pup brought to a veterinarian[1]. How does a veterinarian help a sick pup through its difficulties and eventually to, literally, get on track at its maiden race when it is about 18 months old? Relevance to the question of progress in physics will be brought out gradually, as we work through the chapter.

The vet of our (realistic) story wants to help the sick pup. But, what to do? Understanding a pup's problem, and coming up with a viable medical regime, the vet needs to draw on up-to-date canine science, which includes input from across the sciences[2]. A distinguishing feature of canine science is that it investigates (stages of) canine growth, from zygote to neo-natal puppy, and eventually to maturity.

In physics, instead of a pup story, we have a **PUP** story. Temporarily (for this chapter), I take **PUP** to stand for Physics Under Pressures of history. Like a pup, our **PUP** grows. Our **PUP** also struggles with

---

[1] This is one of McShane's helpful suggestions. See, for example, McShane [2013], pp. 14, 83, 86, 117; and McShane [2011].

[2] Contemporary veterinary science includes bio-mechanics, bio-physics, bio-chemistry, bio-mathematics, animal neuroscience, animal nutrition, zoology, etc. Front line research in veterinary science, of course, goes much further. See, for example, specialties represented in *Research in Veterinary Science*, Elsevier, the journal of the Association for Veterinary Teaching and Research Work.

difficulties such as ongoing confusions and conflicting views, apparent paradoxes, and, at times, has problems obtaining sufficient funding for major initiatives in "big science."[3]

There is, of course, no 'vet' for the physics community in history.[4] Nevertheless, counsel[5] is given and actions are taken, to promote progress and resolve difficulties. But, on what grounds, generally, is such help given? Subtly present are views (vague, descriptive, implicit, or otherwise) about what progress probably is possible, or not, at a given time, in a given situation, and about possible outcomes that should be avoided[6].

Might it not be helpful for the physics community to give some effort to making operative views (of possible progress and decline) explicit?

---

[3] Weinberg [2012].

[4] Philip McShane, though, has used the acronym VET for Vortex Evolutionary Theory. Whatever else McShane means, it was a fresh name for Functional Specialization (http://www.philipmcshane.org/). And, a main message of this book is that functional specialization, VET, will be an effective way to care for our **PUP**.

[5] This asks for only superficial observation. We are touching here on a challenging series of exercises in self-attention, results of which were first reported by Thomas Aquinas, in his *Summa Theologica First Part of the Second Part*, Questions 6–17. See, e.g., Aquinas [1948]. Aquinas' skills in self-attention were remarkable. Although the language of the day did not speak of self-attention, in a series of twelve questions, Aquinas describes twelve main steps leading up to "acts commanded by the will," one of which is "counsel, which precedes choice" (Question 14). The modern scientist may pause over reference to results of a medieval theologian. A base of Aquinas' work, though, was his description of human intellectual acts. The twelve steps are not theological speculation, but are verifiable in one's own experience. Still, being descriptive, his results eventually will need to be connected with an emerging modern neuroscience of deliberation. The twelve steps also can be described in terms of elements of knowing and doing, identified by Lonergan [2001], Appendix A: Two Diagrams, pp. 319–323. See also Figure E.1. For pedagogical introductions, see McShane [1973]; and Benton, Drage and McShane [2005], Chapter 21, pp. 78–82.

[6] Some may suggest: "No, to offer good counsel in the physics community, we neither need, nor generally draw on an understanding of possible progress and decline." Does one have reasons for such a claim? Or, does one suggest that guessing is the best option? If so, why, or why not? And so on. That is, what I invite here is not debate, but self-attention. See note 5. In the philosophic tradition there is the problem of "performative contradiction."

Might it not be practical to grow in an understanding of possible progress-in-actual-contexts, in our spatial and temporal home?

Readers will recall that contemporary philosophy of science already includes a vast literature on scientific development and scientific progress. As already discussed, however, at this stage of history, philosophic traditions tend to emphasize idealized structures remote from scientific experience. There are, for example, ongoing publications regarding Kuhn's assertions about idealized scientific revolutions. Even further from scientific experience is the kind of philosophy of science wherein arguments are

> metaphysical, seeking to give a philosophical account of the most general sorts of things scientists aim to represent – laws of nature, explanations, natural kinds.[7]

This chapter is not the place to study methods in which, for example, "approaches to characterizing scientific progress"[8] are compared and contrasted. Note, also, that I am not trying to imply that the scientific community will not eventually take advantage of the existing literature on method and progress in science[9]. A mature functional collaboration will seek to recycle potentially helpful results in all areas. But, that effective recycling belongs to a distant future. In this chapter, I wish to draw attention to the need and possibility of a seventh main task in physics, one of the eight tasks that will contribute to that future functional cycling. Emergence of, and development in that seventh task will include a growing grasp of ways that our **PUP** will be able to get on functional-track in its own Maiden Race[10].

In Section 7.2, I draw attention to a few examples of explicitly future-oriented work. This is to help us notice a mode of thinking that, in fact, is

---

[7] Bird [1998], p. x.

[8] Bird, [2007], p. 92.

[9] See, e.g., Bird [1998], Chapter 8, Method and progress, pp. 237–287.

[10] How long will it take? As is said of greyhounds, how much "schooling" will be needed, before our **PUP** reaches its Maiden Race in functional tracking? Will it turn out to be about 18 millennia from agricultural beginnings in the Cradle of Civilization? That is, will it be as late as the tenth millennium A. D., when physics (and the rest of the Academy) is on functional-track? The date 9011 A. D. is introduced in McShane [2011a], which includes "a fanciful account of Cosmopolis in 9011 A.D." (McShane [2011a], p. 1).

normal throughout the physics community – in all areas, and in all tasks. But the future-oriented mode is not often discussed. Sections 7.3 and 7.4 invite attention to needed progress in understanding the dynamics of progress and decline. Section 7.5 looks to some of Lonergan's leading ideas[11], doctrinal pointers for how we can climb toward the goal vaguely anticipated in Sections 7.3 and 7.4. Section 7.5 hints at the future possibility of (genetic) sequences of (up-to-date) understandings of our dynamic presence and participation in a dynamic spatial and temporal universe that is *emergent probability*[12].

7.2   Getting in the Mode: 'What to Do?'

Both the fifth and sixth tasks are future-oriented. But, what does that mean, 'future-oriented'? Some authors have asked whether or not looking toward future possibilities is even part of the scientific enterprise. Steve Webb, for several decades a leader in medical physics[13], draws attention to a potential difficulty, at least, when 'looking toward the future' is understood in a naïve sense outside of normal scientific practice. According to Webb, such

> future gazing is a totally unscientific process[14].

While Webb's article is on *medical* physics, some of his comments surely apply to all of physics:

> We cannot know, just as our forebears did not know, the complete course of evolution of … physics. Indeed, many of the key past developments have come by lateral thinking from other fields. Predicting the future is a totally unscientific process. … Nevertheless there can be goals, and key areas in which to work can be identified.[15]

---

[11] Lambert and McShane [2010].
[12] Lonergan [1992], Section 4.2.4.
[13] Ruffle [2008].
[14] Webb [2009], p. 169.
[15] Webb [2009], pp. 174–175.

Prediction, in the sense of working out possible and probable experimental results, is a normal part of science. But, Webb is speaking about prediction of future developments. And what he suggests is confirmed by experience. For, experience in science provides data on scientific process. And, the historical fact is that attempting to 'predict future developments' is not an emphasis in mainstream science. In that sense, too, that is, if we look to scientific practice, 'future gazing' is, as a matter of fact, "unscientific."[16] As Webb points out, however, "goals and key areas" sometimes can be identified.

What, though, might we mean by "goals and key areas"? On what grounds might one area be key for possible progress, while another area not? As throughout this book, this question is about the (actual) physics community. So, let's look briefly to a small sampling of articles that are part of the existing physics literature, articles in which scholars think about what seem to be promising possibilities for development, (from their time) "the future" of physics.

In fact, the last three sections of Webb's article[17] are along such lines. They are, respectively: "The future of medical physics in cancer"; "The future of medical physics in radiotherapy"; and, "Establishing the conditions for good progress." Note too, that his statements in the negative[18] also reveal something in the positive, namely, aspects of Webb's positive understanding of scientific process.

I am just trying to give impressions here, and so I leave it to the reader to go into the details of Webb's article, as well the other articles I reference below. Of course, medical physics is only one part of contemporary physics. So, it will help to also include a few examples from theoretical and experimental physics: "The future of physics"[19]; "Elementary Particle Physics – Where Is It Going?"[20]; "The Future of Physics,"[21]; "Future of

---

[16] See note 14.

[17] Webb [2009], pp. 174–176.

[18] See note 14.

[19] Dyson [1970].

[20] Schopper [1984]. Schopper was a leader of the nuclear physics program at CERN in the 1970's.

[21] Gross [2005].

charm physics" (2012)[22]; "Flavour Physics Beyond the Standard Model: Recent Developments and Future Perspectives"[23]; and "Future flavour physics experiments."[24]

Dyson and Gross speak about active areas of research (in their respective times). With technical precision, both explain what they think are directions worth pursuing. In more recent particle physics, Blanke makes specific recommendations for new experimental work, as do Bigi and Ikaros, and Harnew. Perhaps because future-thinking is not yet well differentiated, the title "The Future of Fundamental Physics" [25] does not in fact well represent the main content of the paper. The paper, instead, offers summaries of various already familiar views and questions. More specifically, the author

> describe(s) the circle of ideas surrounding these questions, as well as some of the theoretical and experimental fronts on which they are being attacked.[26]

The last two sentences of the article, though, do reveal something of the author's lean toward future foundations:

> There must be a new way of thinking about quantum field theories, in which space-time locality is not the star of the show and these remarkable hidden (mathematical) structures are made manifest. Finding this reformulation might be analogous to discovering the least-action formulation of classical physics; by removing spacetime from its primary place in our description of standard physics, we may be in a better position to make the leap to the next theory, where space-time finally ceases to exist.[27]

What does this small sampling of papers tell us about future thinking in physics? Experimental physicists plan new generations of cyclotrons. Theoreticians decide on new directions, on which theories to lay aside, perhaps for possible future development, and which to lay aside as

---

[22] Bigi and Roudeau [2012].

[23] Blanke [2014].

[24] Harnew [2016].

[25] Arkani-Hamed [2012].

[26] Arkani-Hamed [2012], p. 53. At the time of the article by Arkani-Hamed, initial data on the Higgs field had been obtained, but not yet the follow-up confirmation of 2014.

[27] Arkani-Hamed [2012], p. 66.

probably no longer needed. Among other things, these papers provide evidence of what may be obvious to many: whether or not generally adverted to, future-thinking is normal throughout the physics community.

My point here, then, is not that these articles illustrate a functionally specialized task in physics. These examples, however, can help us begin to tune somewhat to future thinking. I also suggest that subtly present in these few examples is, in fact, evidence of the need and possibility of a seventh main task in physics. Making beginnings toward discerning that seventh task will be the work of the rest of this chapter.

7.3 First Impressions: Sensing a Need

Whether it be in efforts to resolve problems, or to strategize future advances, as Section 7.2 shows, future-oriented questions arise. Other examples include: "What do we do to reform physics education?"[28]; "What new efforts should we support through public funding agencies?"; and so on. (Such questions are not limited to physics, but arise throughout the scientific community.) In other words, whether or not adverted to, questions arise, of the form: "In our situation, what are possibilities for progress and for resolving difficulties?"

Whatever one's area of expertise, recommendations can be, and are made. Again, one may (self-) observe that implicit in recommendations intended to promote progress are one's views about what progress is possible, or is not possible (in the near future, in the not too distant future, and in the remote future). In what ways, though, is a particular kind of progress envisaged probable, improbable, probably stable, or not stable? What sequences of combinations of measures and counter-measures are available to one's scientific community that, if implemented, might help the community recover from, or shift beyond, present confusions, or, positively, grow beyond present achievements? What are real capacities and probabilities for growth, development and collaboration in physics, the sciences, the Academy, humanity, in actual contexts, in history?

---

[28] See, for example, Meltzer and Otero [2015], a brief history of physics education in the United States (USA).

7.4   Second Impressions: Genetic Systematics

Section 7.3 ended with vague questions about the need and possibility of understanding something of possible progress and decline in actual contexts. What will such an understanding look like? This book is only a reconnaissance mission, an elementary searching of directions. To explore the question a little, I invite attention to three sources: (1) Chapter 3, which includes a discussion of developments in gauge theory; and of historical understanding; (2) Doctrinal pointings given by Lonergan on *genetic method*[29]; and (3) Our own experience, with us throughout.

Suitably transposed, adding the word 'self,' Lonergan's dense doctrinal paragraph on genetic method[30] provides guidance in the problem of growing in understanding development in physics and science:

---

[29] For the convenience of the reader, I include the full paragraph from *Insight*, on *genetic method*: "Study of an organism begins from the thing-for-us, from the organism exhibited to our senses. A first step is a descriptive differentiation of different parts, and since most of the parts are inside, this descriptive preliminary necessitates dissection or anatomy. A second step consists in the accumulation of insights that relate described parts to organic events, occurrences and operations. By these insights, the parts become known as organs, and the further knowledge constituted by the insights is a grasp of intelligibilities that (1) are immanent in the several parts, (2) refer each part to what it can do and, under determinable conditions, will do, and (3) relate the capacity-for-performance of each part to the capacities-for-performance of the other parts. So physiology follows anatomy. A third step is to effect the transition from the thing-for-us to the thing-itself, from insights that grasp described parts as organs to thing-itself, from insights that grasp described parts as organs to insights that grasp conjugate forms systematizing otherwise coincidental manifolds of chemical and physical process. By this transposition one links physiology with biochemistry and biophysics. To this end, there have to be invented appropriate symbolic images of the relevant chemical and physical process; in these images there have to be grasped by insight the laws of the higher system that account for regularities beyond the range of physics and chemical explanation; from these laws there has to be constructed the flexible circle of schemes of recurrence in which the organism functions; finally, this flexible circle of schemes of recurrence must be coincident with the related set of capacities-for-performance that previously was grasped in sensibly presented organs. ... However, the organism grows and develops. Its higher system at any stage of development not only is integrator but also an operator, that is, it so integrates the underlying manifold as to call forth, by the principles of correspondence and emergence, its own replacement by a more specific and effective integrator" (Lonergan [1992], 489–490).

[30] See note 29.

(Self-) study of an organism begins from the thing-for-us, from the (self-) organism exhibited to our senses. A first step is a descriptive differentiation. ...[31]

No doubt, an up-to-date theory of possible progress needs to be partly informed by growth and development that already occurred, that (even if only provisionally) already has been identified. As we have seen in Chapter 3, for example, the analysis given by O'Raifeartaigh provides details on sequences of gauge theories that partly were advances but also partly carried forward ongoing confusions[32]. In other words, the historical development of gauge field theory, leading up to what is now called the Standard Model, has been through what can be called *genetic-dialectic sequences of systems.*[33]

Where O'Raifeartaigh identified particular sequences in the emergence of gauge field theory, Lonergan inquired into method in historical understanding. Already quoted in full in Chapter 3, one of his observations about historical understanding invites us to notice that

> successive systems which have been progressively developed over a period of time have to be understood.[34]

With the results of both O'Raifeartaigh and Lonergan in mind, the need for attending to (genetic-dialectic) sequences becomes increasingly evident. However, the difficulty of the problem grows enormously when we recall that no area in physics (or any part of science) is an island[35]. And

---

[31] I am referring to a future stage of scientific development.

[32] We see some main sequence terms represented in Straumann [1996], Figure 1, p. 4.

[33] For now, the name 'genetic-dialectic' is at least descriptive of actual sequences. See, however, note 43. Explanatory accounts will emerge within future Genetic Systematics. See also note 38. Similarly, for now, the name 'system' also is merely descriptive. In preliminary description it is evident already that actual many-disciplinary science consists of complex "layerings" of both descriptive and explanatory elements. In a future Genetic Systematics, the meaning of 'system' will luminously embrace, but will not be limited to, systems in symbolic and formal logic, structures in Kuhnian analysis, and systems as defined in contemporary systems theories.

[34] Lonergan [2013], 175–177.

[35] Contemporary reflection speaks of inter-disciplinary and multi-disciplinary science. Whatever one's philosophy of space and time, however, lengths and times, symbolizations and expressions all are human experience. Even when one's focus is on, say, understanding

so, while his focus is on physics, Ryde's observation about the development of ideas in physics can be understood to hold for all of science:

> The development of ideas cannot be treated in isolation; progress in one sphere has often furthered the evolution in another, adjacent one. Sometimes ideas also from a certain domain of physics have been applicable to a totally different one.[36]

More accurate description, though, is in terms of genetic-dialectic sequences emerging in the ongoing history of physics and science.

I have been referring to Chapter 3, on functional history. In the context of the present chapter, what is taking shape is a further significance of Lonergan's insight about historical understanding[37]. For, besides asking about genetic-dialectic sequences that already have occurred, there is *Genetic Systematics*[38]. That is, we also can ask about (really) possible[39] genetic-dialectic sequences of systems in physics and science, whether or not they already have occurred. Think, now, of helping our **PUP** struggling and growing in history. In other words, a major task is called for, a Genetic Systematics that would, in particular, take advantage of the cumulative results of the first six specialties[40].

---

the geometry of elementary particles, explaining that focus and understanding needs an omni-disciplinary context that includes human development, consciousness and intellect. See note 31.

[36] Ryde [1994], p. 5.

[37] Lonergan [2013], 175–177.

[38] See, e.g., McShane [2013], 84–85. Note that a Genetic Systematics will include the contra-factual. There is contra-factual history, grasped relative to one's view of what is possible. But, the contra-factual in Genetic Systematics will be part of an up-to-date Genetic Systematics. And so, mere speculation about plausible lines of development will not necessarily contribute to the new Genetic Systematics. Aspects of the existing literature (e.g., in Pessoa Jr. [2001] and Hawthorn [1991]) eventually will find new ordering within functional control of meaning. Some aspects will be identified as contra-factual and contributive to functional history. Some will find a home in a Genetic Systematics. Some will be identified as conceptualist speculation.

[39] 'All (really) possible'? A practical theory of progress and decline will not to be about any and all conceivable sequences in any and all possible universes, but rather will be about really possible sequences in history.

[40] See also, the last paragraph of this chapter.

7.5 Lonergan's *Sensei*[41]-able Leads

An up-to-date Genetic Systematics is well beyond the reach of present achievement. For now, I invite attention to a lead given as part of Lonergan's later introduction to a *canon of explanation*: (Note that text is already recorded in Chapter 2 (notes 47 and 48) but, in this new context, is repeated here for convenience of the reader.)

> The explanatory differentiation of the protean notion of being involves three elements. First, there is the genetic sequence in which insights gradually are accumulated by man. Secondly, there are the dialectical alternatives in which accumulated insights are formulated, with positions inviting further development and counterpositions shifting their ground to avoid the reversal they demand. Thirdly, with the advance of culture and of effective education, there arises the possibility of the differentiation and specialization of modes of expression; and since this development conditions not only the exact communication of insights but also the discoverer's own grasp of his[42] discovery, since such grasp and its exact communication intimately are connected with the advance of positions and the reversal of counterpositions, the three elements in the explanatory differentiation of the protean notion of being fuse into a single explanation.[43]

There is no doubt that following up in these doctrinal pointers will be a community achievement of the distant future. But, we now have an accumulation of clues, hintings that Genetic Systematics will seek best-to-date understanding of what elsewhere was named *the concrete intelligibility of Space and Time*[44]. By the same token, we get glimpsings of a seventh task, namely, *functional systematics*. Not only can we dream

---

[41] *Sensei*: Lonergan as *teacher*.

[42] Using the male pronoun was a sign of the times as well as of Lonergan's biographical context. But, grammatical idiosyncrasy does not imply bias, as is revealed in "a close and careful interpretation of Lonergan's intended meaning in the light of his entire corpus" (Crysdale [1994], p. 7).

[43] In various writings, McShane has named the paragraph simply *60910*. For McShane, "it stands out in the book (*Insight*, Lonergan [1992]) as a central now-impossible challenge of the Tower Enterprise" (McShane [2013], p. 78). The Tower will be an implemented functional collaboration. See Lambert and McShane [2010], The Tower of Able: Lonergan's Dream, p. 163.

[44] Lonergan [1992], The Concrete Intelligibility of Space and Time, sec. 5.5, pp. 194–195. But, see also Quinn [2016].

of a final theory in physics[45], but we can think, precisely, luminously and explanatorily, of genetic-dialectic sequences of *emergent probability*[46].

How, though, can we take advantage of results of functional systematics? As the question suggests, there is at least one more main task, discussion of which belongs to the next chapter.

7.6 References

Aquinas, T. (1948) *St. Thomas Aquinas, Summa Theologica, Complete Edition in Five Volumes*, trans. by Fathers of the Dominican Province (Christian Classics, Notre Dame, IN).

Arkani-Hamed, N. (2012) The Future of Fundamental Physics, *Daedalus, Proc. Am. Acad. Arts Sci.*, 141 (3), pp. 53–66.

Benton, J., Drage, A., and McShane, P. (2005) *Introducing Critical Thinking* (Axial Publishing, Vancouver).

Bigi, I. I. and Roudeau, P. (2012) Future of Charm Physics, *C. R. Phys.*, 13, pp. 133–140.

Bird, A. (1998) *Philosophy of Science* (McGill-Queen's University Press, Montreal, QC and Kingston, ON).

Bird, A. (2007) What is Scientific Progress? *NOÛS* 14:1, pp. 92–117.

Blanke, M. (2014) Flavour Physics Beyond the Standard Model, CERN-PH-TH-2014-240 TTP14-033, December 3, presented at *The 8th International Workshop on the CKM Unitarity Triangle* (CKM 2014), Vienna, Austria, September 8–12, 2014.

Crysdale, C. S. W., ed. (1994) *Lonergan and Feminism*, 2nd edition (University of Toronto Press, Toronto).

Dyson, F. J. (1970) The Future of Physics, *Phys. Today* 23 (9), http://dx.doi.org/10.1063/1.3022330.

Gross, D. (2005) The Future of Physics, *Int. J. Mod. Phys. A*, vol. 20, no. 26, pp. 5897–5909.

Harnew, N. (2016) Future flavour physics experiments, *Ann. Phys.* (Berlin) 528, No.1–2, pp. 102–107.

Hawthorn, G. (1991) Plausible Worlds, Possibility and Understanding in History and the Social Sciences (Cambridge University Press, Cambridge).

Lambert, P. and McShane, P. (2010) *Bernard Lonergan. His Life and Leading Ideas* (Axial Publishing, Vancouver).

Lonergan, B. (1975) *Method in Theology* (Darton, Longman & Todd, London).

Lonergan, B. (1985) *A Third Collection, Papers by Bernard J. F. Lonergan, S. J.*, ed., Crowe, F.E., S. J. (Paulist Press, New York).

---

[45] Weinberg [1994].
[46] Lonergan [1992].

Lonergan, B. (1992) *Insight: A Study of Human Understanding*, eds., Crowe, F. E. and Doran, R. F., vol. 3 of *Collected Works of Bernard Lonergan* (University of Toronto Press, Toronto).

Lonergan, B. (2001) *Phenomenology and Logic: The Boston Lectures on Mathematical Logic and Existentialism*, ed., McShane, P. J., vol. 18 in *The Collected Works of Bernard Lonergan* (University of Toronto Press, Toronto).

Lonergan, B. (2013) *Early Works on Theological Method 2*, trans. Michael G. Shields, eds., Doran, R. M., and Monsour, H. D., *Collected Works of Bernard Lonergan*, vol. 23 (Toronto: University of Toronto Press).

McShane, P. (1970) *Randomness, Statistics and Emergence* (University of Notre Dame Press, Notre Dame, IN).

McShane, P. (1973) *Wealth of Self and Wealth of Nations* (Exposition Press, New York).

McShane, P. (1976) *The Shaping of the Foundations: Being at home in the Transcendental Method* (University Press of America, Washington, D. C.).

McShane, P. (1980) *Lonergan's Challenge to the University and the Economy* (University Press of America, Washington, D. C.).

McShane, P. (1998) *Economics for Everyone* (Axial Publishing, Vancouver).

McShane, P. (2008) Metagrams and Metaphysics, *Prehumous* 2 http://www.philipmcshane. org/prehumus/.

McShane, P. (2007) The Importance of Rescuing *Insight*, ch. 12 in Liptay, J. J. Jr. and Liptay, D.S. (2007), *The Importance of Insight, Essays in Honour of Michael Vertin* (University of Toronto Press, Toronto), pp. 199–225.

McShane, P. (2011a) Arriving in Cosmopolis, Keynote Address at *The First Latin American Lonergan Workshop*, Puebla, Mexico, June, 2011, http://www. philipmcshane.org/website-articles/.

McShane, P. (2011b) The Future of Functional History, *FuSe* 15, http://www. philipmcshane.org/fuse/.

McShane, P. (2013) *Futurology Express* (Axial Publishing, Vancouver, Canada).

Meltzer, D. E. and Otero, V. K. (2015) A brief history of physics education in the United States, *Am. J. Phys.*, 83, 447–458.

Pessoa Jr., O. (2000) Counterfactual Histories: The Beginning of Quantum Mechanics, Philosophy of Science, in vol. 68, No. 3, *Supplement: Proceedings of the 2000 Biennial Meeting of the Philosophy of Science Association. Part I: Contributed Papers* (September) (The University of Chicago Press, Chicago), pp. 519–530.

Quinn, T. (2016) Interpreting Lonergan's 5th Chapter of *Insight*, ch. 2 in Brown, P. and Duffy, J., eds., *Seeding Global Collaboration* (Axial Publishing, Vancouver), pp. 29–44.

Quinn, T. (2017) *Invitation to Generalized Empirical Method* (World Scientific Publishing, Singapore).

Ruffle, J. (2008) Steve Webb: a life in a day, *Medical Physics Web* (Institute of Physics Publishing, Bristol).

Ryde, N. (1994) *Development of Ideas in Physics* (Almqvist and Wiksell International, Stockholm, Sweden).

Schopper, H. (1984) Elementary Particle Physics – Where is it going? *Nucl. Inst. Meth. Phys. Res.* (219), pp. 54–65.

Smolin, L. (2006) The Trouble with Physics, The Rise of String Theory, the Fall of a Science, and What Comes Next (Houghton Mifflin, NY).

Straumann, N. (1996) Early History of Gauge Theories and Weak Interactions, Invited talk at the *PSI Summer School on Physics with Neutrinos*, Zuoz, Switzerland, August 4–10, 1996. https://archive.org/details/arxiv-hep-ph9609230.

Webb, D. (2009) The contribution, history, impact and future of physics in medicine, *Acta Oncol.*, 48, pp. 169–177.

Weinberg, S. (2012) The Crisis of Big Science, *The New York Times Review of Books*, May 10.

Weinberg, S. (1994) *Dreams of a Final Theory* (Vintage Books, NY).

Chapter 8

Functional Communications

8.1 Introduction

An emphasis of functional systematics is conceiving possibilities. What, though, of local needs? The first paragraph of the *Magna Moralia* (of the Aristotelian tradition) reveals something of the problem:

> Since our purpose is to speak about matters to do with character, we must first inquire of what character is a branch. To speak concisely, then, it would seem to be a branch of nothing else than statecraft. For it is not possible to act at all in affairs of the state unless one is of a certain kind, to wit, good. Now to be good is to possess the excellences. If therefore one is to act successfully in affairs of state, one must possess good character. The treatment of character, then, is, as it seems, a branch and starting point of statecraft. And as a whole it seems to me that the subject ought rightly to be called, not Ethics, but Politics.[1]

So far in this book, seven main tasks have been described. This chapter draws attention to an eighth, *functional communications*.

> It is a major concern, for it is in this final stage that (omni-disciplinary) reflection bears fruit. Without the first seven stages, of course, there is no fruit to be borne. But without the last the first seven are in vain, for they fail to mature.[2]

Where functional research will be alert to anomalies (positive or negative) in labs, literatures, locales and communities, the eighth

---

[1] Barnes [1984], vol. 2, p. 1868.

[2] Lonergan [1975], p. 355. In the quotation, the word 'omni-disciplinary' replaces 'theological' from the original text. As mentioned in the Preface, and as being brought out as we go, functional collaboration will be omni-disciplinary.

functional specialty will mediate local needs. Among other things, it, or rather they (*characters* of the eighth task) will draw creatively on results of seven prior functional specialties, seven prior *situation rooms*.

Section 8.2 reveals something of the gap between present contexts of concern and actual contexts. Most of the examples mentioned relate to physics. But, keep in mind that an eighth functional specialty will be distinguished not by results of investigations or sources of data, but by being an eighth stage in

the process from data to results.[3]

Section 8.3 is on the problem of meeting local needs, and points to a few of the more obvious problems of present times, in classrooms, offices, laboratories, professional communities, towns and cities.

Section 8.4 brings a few aspects of the eighth task into shallow relief.

In Section 8.5, two diagrams are introduced, both invented by Philip McShane.[4] With the help of the two diagrams, the possibility of a subtle and verifiable heuristics begins to emerge: Communications C_{ij} will generate flows in history. The global topology of functional communications and dependencies will be a World Markov Process. Precision in identifications, however, necessarily will be the work of future generations and generators.

8.2   Actual Contexts: Physics and Societies

There have been various characters in physics and science known for their concern for societies and cultures. For instance, there was the "polymathic"[5] Victor Weisskopf (1908–2002) who was

regarded as one of the great men of 20th century physics.[6]

---

[3] See note 2. See also, Lonergan [1975], p. 125.
[4] See, for example, McShane [1998a], pp. 108–109.
[5] Stefan [1998], p. v.
[6] Stefan [1998], p. v.

In addition to his contributions to nuclear physics and quantum electrodynamics, he also was known for his career-long concern for

the social and political responsibilities of science and scientists. ... (H)e was called the conscience of the physics community; .... (and later) the conscience of the entire international scientific community.[7]

Weisskopf wrote of the

privilege of being a scientist. ... (the) privilege to be deeply concerned with the involvement of science in the events of the day. ... The (scientist) cannot avoid being drawn in one form or another into the decision-making process regarding the applications of science, be it on the military or on the industrial scene. (The scientist) may have to help, advise, or to protest, whatever the case may be. There are different ways in which (the scientist) will get involved in public affairs: (the scientist) may address (themselves) to the public when (they) feel that science has been misused or falsely applied; (the scientist) may work with (their) government on the manner of application of scientific results to military or social problems.[8]

Of course, not all scientists can be expected to be directly involved in all "scenes."[9] Nevertheless, Weisskopf was speaking about real needs and actual contexts. And, while indirectly so, his comments touch on the eighth main task.

8.3 Local Needs

To notice something of the presence and need of the further task,

we will have to look deeper into the question of the relevance of science to society as a whole.[10]

Not infrequently,

the involvement of science in the events of the day[11]

---

[7] Stefan [1998], p. v.

[8] Weisskopf [2003], p. 51. The male pronouns in the original text are replaced by the more generic 'scientist,' 'them,' and 'they.'

[9] "The (scientist) cannot avoid being drawn in one form or another into the decision-making process regarding the applications of science, be it on the military or on the industrial scene." See note 8.

[10] Weisskopf [2003], p. 48.

[11] See note 8.

has led to serious difficulties. An all too familiar example is that

> our advancing knowledge ... (has given us) powerful and dangerous means of destruction.[12]

In a preliminary attempt to get a hold of this problem, Weisskopf pointed out that

> (c)ompassion without knowledge is ineffective; and knowledge without compassion is inhumane.[13]

But, methodology was not his area of expertise. He had to leave the solution of the problem to other scholars. Late in his career, Weisskopf had to settle for:

> The paradox is the problem we live with.[14]

To glimpse something of the full complexity of "the problem we live with,"[15] let's look now to a few examples. Physics and medicine have been working together since ancient times. Medieval meanings of the word *fysike* include "the art of healing." Advances in modern medical physics are well-known[16]; and, the ongoing collaboration is expected to be a "long and happy marriage."[17] But, nuclear physics in modern medicine also goes into the design of nuclear warheads that are being tested in oceans and mountains. And, there are now established literatures devoted to mathematical modeling of nuclear testing, political conflict, violence, war and terrorism[18].

---

[12] Stefan [1998], p. 77.

[13] Attributed to Victor Weisskopf. See, for example, Stefan [1998], p. v.

[14] This is from a speech given by Weisskopf at Los Alamos, late in his career. See Stefan [1998], p. 77.

[15] Note 14.

[16] See, for example, Benefits of HEP, http://science.energy.gov/hep/benefits-of-hep/. Note: 'HEP' stands for High Energy Physics. See also Mielczarek [2006], Keevil [2012] and O'Shea [2012].

[17] Knight [2012].

[18] Glaser and Mian [2008]; and Morgenstern *et al.* [2013].

Perhaps the meeting of physics and society will seem less problematic if we look to non-military examples. For instance, fluid dynamics, electrical engineering, the science of materials, and many other areas of applied physics and chemistry all are essential in the design, construction and operation of modern dams, irrigation systems for farmlands, as well as water systems for towns and cities.[19]

In many parts of the world, there are *large-scale dams*. At present, there are at least 50 such dams in India. But, there are also plans underway in India, Pakistan, Nepal and Bhutan to build several hundred new large-scale dams in the Himalayas[20]. No doubt, we need water; electricity is convenient; and flood control has been helpful. Also, human populations continue to increase, as does the need for water, electricity and arable land. But, will it be wise to build so many large-scale dams in the Himalayas?

The rivers that originate in the Himalayas provide sustenance, livelihoods and prosperity to millions of people living in a vast area that stretches from the Indus Basin plains of Pakistan in the west to Bangladesh in the east.[21]

At the same time, among known effects of the proposed large-scale dams, they will: destroy rivers; eliminate riparian and littoral communities and their economies[22]; cause extensive damage to eco-systems on a continental scale; significantly contribute to global warming[23]; and significantly reduce flux in the world's hydrological cycles[24].

What rationale could justify plans that will cause such damage to peoples and planet? Partly, damages are considered justifiable by appealing to Establishment Economics. Among its various principles, Establishment Economics promotes ongoing 'growth' and 'maximization

---

[19] See, for example, Green [2007].

[20] Plans for 100's of new dams are underway world-wide. See, e.g., Ansar at al [2014].

[21] https://www.internationalrivers.org/files/attached-files/ir_himalayas.pdf.

[22] See note 21.

[23] Recent results would seem to be conclusive, that reservoir waters are major contributors of global greenhouse gases (Deemer and Harrison [2016]).

[24] The problem, in fact, is global. The growing literature is extensive. See, for example, Wang et al [2006], Charlton [2013], Pearsona et al. [2016], Gumiero et al. [2013].

of assets' (where 'growth,' 'maximization' and 'assets' all are defined within Establishment Economics).[25]

This is not to suggest that contemporary economists intend harm, as such.[26] What is in evidence, however, is that traditional *ad hoc* combinations of scientific advance, contemporary economic speculation and education are inimical to local needs, to the survival of cultures, societies and ecologies.[27] Evidently, an often quoted statement of Bernard Lonergan still applies[28]:

> (P)hilosophers for at least two centuries, through doctrines on politics, economics, education, and through ever further doctrines, have been trying to remake man, and have done not a little to make human life unlivable.[29]

8.4 Executive Reflection[30]

How can we attend better to local needs? Some hints are available in the questions posed by Dorothy Green (1929–2008). Green was an environmentalist who was interested in, among other things, responsible management of water resources. In the quotation below, we can observe certain features of her inquiry. We can replace words like 'water,' 'land,' ..., with x, y, ... To illustrate the point, in the following quotation I have inserted x and y (in parentheses) beside the corresponding words 'water' and 'land':

---

[25] What is the alternative to Establishment Economics? In fact, present economic ills brought about through establishment economics have their roots in errors identified more than 70 years ago, by Bernard Lonergan. See Lonergan [1998]. Introductions to Lonergan's economics and supporting literature include: McShane [1998b], [2002], [2010]; and [2010], ed.; [2013], ch. 7; [2014]; McShane and Anderson [2002]; Anderson [2012]; Shute [2010a] and Shute [2010b]. See also Chapter 9, *D4a* and *D4b*. In the new economics, there will be economists in every village, town and neighborhood.

[26] See, e.g., the new Springer journal, *Ethical Economy. Studies in Economic Ethics and Philosophy.*

[27] Establishment economics obviously does not look to local needs. Nor is it explanatory of economic processes.

[28] For the present context, I add the phrase in italics.

[29] Taken from Lonergan's 1959 lectures on education. See Lonergan [1993], p. 232.

[30] McShane [1998a], p. 109; and McShane [2013], p. 93.

As the issues are explored, here are some of the many questions that must be addressed. What are the current decision-making processes? How and by whom are the decisions about water (x) use made? How and by whom should they be made? Are these decisions responsive to changing needs? Should they be left to our elected representatives? At which level of government – the state legislature? city councils and planning commissions? water (x) agency boards? professional planners? Land (y) use decisions such as where to permit development and how the land (y) is developed impact our water (x) supply. What role should water (x) agencies play in land (y) use decisions? What role should there be for citizen involvement in these difficult and important decisions? [31]

Notice that problems are concrete. Of course, think too of the examples described in Section 8.3. There are *local* needs (where *local* is not necessarily merely geographic, but regards the topology of global dependencies and human collaboration).

What are possible ways forward? What are combinations and orderings that might contribute to progress and health of a dynamic community? Such questions arise, whether individuals be in a state legislature, city council, a planning commission or agency, in the arts or sciences, fishing or farming, and so on, as the case may be.

Is there not an evident need for collaborative reflection on local possibilities that are

responsive to changing needs? [32]

But, results of such reflection need to reach communities, peoples and groups of different views, and eventually be implemented.

What we find here are hintings of two aspects of an eighth main task in the Academy. There is need of local collaborative reflection that selects creatively from ordered ranges of possibilities. At the same time, there is also need for results to be communicated – to diverse groups in the world, toward the possibility of implementation.

To end this section, let's explore something of what this will look like when functional collaboration has become operative. Functional collaboration will include the seven functional specialties already discussed in this book. But, in order to respond effectively to changing

---

[31] Green [2007], p. 2.
[32] See note 31.

local needs of communities, a further creative effort will be needed. A highpoint in the Academic Enterprise, then, will be an eighth main task that will draw on, in particular, an up-to-date Genetic Systematics. The eighth task will be an Executive Reflection that mediates

> between ... the genetic sequences of empirically grounded understandings of the new systematics, and the varieties of disciplines, cultures, media, ... [33]

Executive Reflection will be (as Weisskopf realized was necessary), both compassionate and knowledgeable. It also will be discerning; sensitive to; open to; creative, interested in, and understanding of, local differences and local needs, possibilities and growth potentials. Progress in Executive Reflection will include development in being able to luminously communicate with diverse groups – scientists, scholars, artists, educators, economists, religious traditions, groups of places and times, aesthetics and sensibilities[34], habits and mentalities. Able to take advantage of the results of seven prior functional specialties, in the eighth functional specialty

> (scholars will) learn to talk to the rest of the world. (The Academy) has to explain how one is to communicate to all peoples, all cultures, all classes. And further, it

---

[33] McShane [1988].

[34] It might, for instance, take up and develop results on "warmth" in varieties of expression. (See, e.g., Aaker, Stayman and Hagerty [1986].) Note, too, that there are languages in Physics Education (Riendeau [2014]), revealed, for example, in ongoing efforts to make progress in learning how to teach thermodynamics in introductory physics, chemistry and biology (Alonso and Finn [1995], Reif [1999], Lee [2001], Dreyfus et al. [2015]). But, there are also languages in all areas of physics and other sciences; in governments' Departments of Energy; in funding agencies and politics (see, for example, Burton, [2009]). There are languages and mentalities of diverse applications of fluid dynamics, electrical engineering, civil engineering and hydro-electrical engineering. Languages and mentalities of corporate fishing industries of Alaska (USA) are different from languages and mentalities of small business fishing groups in Kerala (India). There will be languages and mentalities of local economies. (Again, in the new economics, there will be an economist in every town. See references in note 25.) Executive Reflection will have something to offer nations in efforts to work and live together. For example, there would be advantages in having Executive Reflection presence at world conferences on climate change, energy, culture, economics and water management.

has to be able to effect the transpositions necessary to do this and to use creatively the diverse media of communication that exist today[35].

What is being envisaged, then, is an eighth functional specialty where the *per se* task is

dialogue with other views[36].

The chapter started with a quotation regarding 'character' (from the Aristotelian tradition). Characters of functional communications will contribute to an eighth stage in *functional global caring*, normatively an eightfold process from data to results[37].

8.5 Communications in World Enterprise

This section introduces two diagrams invented by Philip McShane[38]. They help in our heuristics of communications being envisaged.

As previous chapters of the book suggest, results of functional specialization normatively will be cumulative, through forward-communications *C*(Research, Interpretation), *C*(Interpretation, History), ..., *C*(Systematics, Communications). For the present context, let's use *C9* to represent communications in the large zone of meaning beyond the central cycling of functional collaboration. It is too soon for us to reach precision here. In particular, we need to leave to future scholars the challenging work of identifying genera and species, boundary layerings, mergings and emergings. With that proviso, we are ready now for the first diagram of this section, a diagram that highlights the *care-full* climb through the functional specialties:

---

[35] Lonergan [2010], p. 453 (where 'today' is a relative term).
[36] McShane [2007], p. 203.
[37] See note 3.
[38] See, for example, McShane [1998a], pp. 108–109.

"Executive Reflection mediates between...
the genetic sequence of empirically-
grounded understanding of the new
systematics and the varieties of disciplines,
cultures, media..." (McShane [1986], 151).

"a succession of transpositions to ever more
determinate contexts" (Lonergan [1975], 142).

"not a static pool of information but
a moving stream of cumulative and
progressive information" (Lonergan [1985], p. 82).

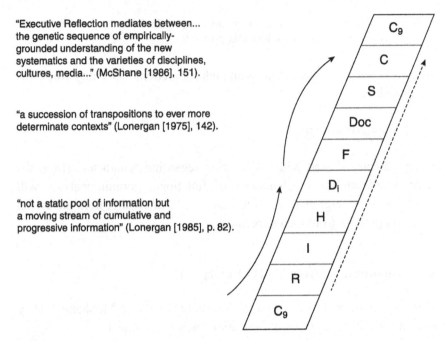

Figure 8.1 Central Functional Climb. Stairway Diagram.[39]

Figure 8.1 includes *C9*. But, as already mentioned, there also will be internal communications *C*(Research, Interpretation), *C*(Interpretation, History), and so on.[40] These can be more concisely written C_{12}, C_{23}, ..., C_{78}, respectively.

Of course, other cases also will be normal. A kind of reverse flow will be of special significance[41]. In Section 8.3, for example, it is observed that

---

[39] For more details, diagram on page 109, of McShane [1998a], from which Figure 8.1 was designed. Note, too, that I have added the quotation from Lonergan [1975], p. 142.

[40] In four centuries, physics has made notable progress in methods of (pre-functional) communication between experimental physics and theoretical physics. In a similar way, working out how to communicate functionally will be part of a gradual development, something to be learned as we go.

[41] In several of his recent works, Philip McShane speaks of 'backfire.' Hints also are provided by Bernard Lonergan, where he describes the interdependence of the functional specialties (Lonergan [1975], pp. 138–144; and Lonergan [2010], pp. 462–466). In particular, Lonergan points out that the first phase also depends on the second. However, "the greatest care must be exercised" (Lonergan [2010], p. 463). I would not presume to

there are problems with plans to build hundreds of new large-scale dams. Initially, efforts to meet local needs are through conversations among, for example, financiers, engineers, consultants in environmental science, politicians, journalists, neighborhood councils and (in future, local[42]) economists. Are groups to follow through with what might be expedient in the short-term? Or, are there other solutions that would better meet global-local and long-term needs?

The problem may at first be beyond *C9* reckoning. Might communications with those in functional communications help? But, the problem may similarly challenge present expertise of functional communications. The problem might then be shared with functional systematics. Indeed, a problem might be so challenging as to be repeatedly pushed back to prior functional specialties. Communications, C_{87}, C_{76}, ... may eventually lead to C_{21}. If functional research is able to helpfully identify anomalies, the forward cycle can begin anew. Eventually, the Academy and communities can return to the problem of meeting local needs, but with new back-up from the entire functional cycle.

The cycling and reverse-cycling described above will be normative. But, these two flows neither exhaust the possibilities nor imply restrictions on what might be helpful. Cross-over communications already occur in *ad hoc* pre-functional ways. We can expect internal communications between *all* functional specialties.[43] To include all possibilities, we can write C_{ij}, *i*, *j = 1, ..., 8.*

---

think that my preliminary grasp and discussion of the problem here is comparable to the subtleties reached by the accelerating McShane, nor to the nuanced view attained by the genius Lonergan. However, I do suggest that my superficial description of reverse communications is illustrative of normative possibilities.

[42] Local economists will have an important say in such matters. See references in note 25.

[43] For example, a foundations person seeking to improve heuristics about potencies, might be helped by new (functional) research regarding anomalies in measurements and speculations about "dark energy" and "dark matter." Recall, however, Lonergan's cautionary remark. (See note 41.) The tract continues: "There is (for instance) a dependence of the first phase on the second. But undue influence absolutely has to be avoided. If there were undue influence, the second phase would become independent of the first. It would be dictating to the first what it should be. And it becomes isolated and sterile if it does that. Its life is out of the data, out of the first phase. And if it starts acting too strongly back

In mature functional collaboration, internal communications C_{ij} will be relatively luminous, with a control of meaning at the level of the times. The entire Academy moves forward in time, symbolized here simply by t^+. Bringing all of this together in a single diagram, we can begin to see that the global dynamics of functional collaboration will be a Global Markov Process:

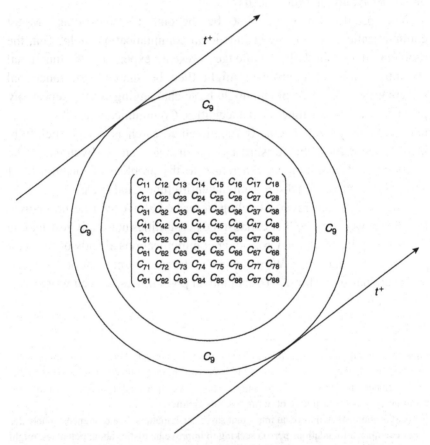

Figure 8.2. Functional Collaboration: A Global Markov Process[44]

---

again, its corrupting the first phase, corrupting the challenge that comes to it from the first phase" (Lonergan [2010], p. 463). See also Lonergan [1975], p. 143.

[44] See McShane [1998a], p. 108. See also ch. 5, note 23.

8.6 References

Aaker, D. A., Stayman, D. M. and Hagerty, M. R. (1986), Warmth in Advertising: Measurement, Impact and Sequence Effects, *J. Cons. Res.* Vol. 12, March, pp. 365–381.

Alonso, M. and Finn, E. J. (1995) An Integrated Approach to Thermodynamics in the Introductory Physics Course, *Phys. Teach.*, vol. 33, May, pp. 296–310.

Anderson, B. (2012) Is there anything special about business ethics? *J. Macrodyn. Anal.*, vol. 7, pp. 454-496.

Ansara, A., Flyvbjergb, B, Budzierb, A. and Lunnc, D. (2014) Should we build more large dams? The actual costs of hydropower megaproject development, Energy Policy 69, pp. 43–56.

Barletta, W. A. (2012) Resource Letter AFHEP-1: Accelerators for the Future of High-Energy Physics, in *Am. J. Phys.*, vol. 80, issue 2, Feb., pp. 102–112.

Barnes, J. (1984), ed., *The Complete Works of Aristotle* (Princeton University Press, Princeton).

Burton, H. (2009) *First Principles, The Crazy Business of Doing Serious Science* (Key Porter Bools, Toronto).

Charlton, L. (2013) China's Great Dam Boom: An Assault on its Rivers, *Environment 360*, http://e360.yale.edu/feature/chinas_great_dam_boom_an_assault_on_its_river_syst ems/2706/.

Colgan, J. D. (2013) *Petro-Aggression: When Oil Causes War* (Cambridge University Press, New York).

Deemer, M. and Harrison, J. (2016) Greenhouse Gas Emissions from Reservoir Water Surfaces: A New Global Synthesis, *Bioscience*, Oct., vol. 66, Issue 11, pp. 949–964.

Dreyfus, B. W., Geller, B. D., Meltzer, D. W. and Sawtelle, V. (2015) Resource Letter TTSM-1: Teaching Thermodynamics and Statistical Mechanics in Introductory Physics, Chemistry and Biology, *Am. J. Phys.*, vol. 83, issue 1, Jan., pp. 5–21.

El-Affendi, A. ed. (2015) *Genocidal Nightmares, Narratives of Insecurity and the Logic of Mass Atrocities* (Bloomsbury, New York)

Glaser, G. and Mian, Z. (2008) Resource Letter PSNAC-1: Physics and society: Nuclear arms control, in *Am. J. Phys.*, vol. 76, issue 1, Jan., pp. 5–14.

Green, D. (2007) *Managing Water, Avoiding Crisis in California* (University of California's Press, Berkeley).

Gumiero, B., Mant, J., Hein, T., Elso, J. and Boz, B. (2013) Efforts in Europe, to reverse course, challenges and possibilities: Linking the restoration of rivers and riparian zones/wetlands in Europe: Sharing knowledge through case studies, *Ecol. Eng.*, vol. 56, July, pp. 36–50.

Hobson, A. (2007) Resource Letter PSEn-1: Physics and society: Energy, in *Am. J. Phys.*, vol. 75, issue 4, April, pp. 294–308.

Keevil, S. F. (2012) Physics and Medicine 1, Physics and medicine: a historical perspective, *The Lancet*, vol. 379, April 21, pp. 1517–1524.

Knight, P. (2012) Physics and medicine – two tips for a long and happy marriage, in *The Lancet*, vol. 379, April 21, pp. 1463–1464.

Lee, K-C. (2001) How to each statistical physics in an introductory physics course, *Am. J. Phys.*, vol. 69, issue 1, Jan., pp. 68–75.

Lonergan, B. J. F. (1975) *Method in Theology* (Darton, Longman and Todd, London).

Lonergan, B. J. F. (1985) *A Third Collection*, ed. Crowe, F. E. (Paulist Press, New York/Mahwah)

Lonergan, B. J. F. (1993) *Topics in Education, The Cincinnati Lectures of 1959 on the Philosophy of Education*, eds., Doran, R. M. and Crowe, F. E., vol. 10 of *The Collected Works of Bernard Lonergan* (University of Toronto Press, Toronto).

Lonergan, B. J. F. (1998) *For a New Political Economy*, ed. by Philip McShane, *Collected Works of Bernard Lonergan*, vol. 21 (University of Toronto Press, Toronto).

Lonergan, B. J. F. (1992) *Insight: A Study of Human Understanding*, eds., Crowe, F. E. and Doran, R. M., *Collected Works of Bernard Lonergan*, vol. 2 (University of Toronto Press, Toronto).

Lonergan, B. (2010) *Early Works on Theological Method 1*, vol. 22 in Doran, R. M. and Croken, R. C. eds., *Collected Works of Bernard Lonergan* (University of Toronto Press, Toronto).

McShane, P. (1988) Systematics, Communications, Actual Contexts, in, Riley, B. and Fallon, T., eds., *Religion in Context* (University Press of America, Lanham, MD), pp. 59–86.

McShane, P. (1998a) *A Brief History of Tongue, From Big Bang to Coloured Wholes* (Axial Publishing, Vancouver).

McShane P. (1998b) *Economics for Everyone: Das Jus Kapital* (Axial Publishing, Vancouver)

McShane, P. (2002) *Pastkeynes Pastmodern Economics: A Fresh Pragmatism* (Axial Publishing, Vancouver).

McShane, P. (2007) The Importance of Rescuing *Insight*, in Liptay J. J. Jr. and Liptay D. S. eds., *The Importance of Insight. Essays in Honour of Michael Vertin* (University of Toronto Press, Toronto), pp. 199–225.

McShane, P., ed. (2010) *Do You Want a Sane Global Economy? Special Issue, Divyadaan, J. Phil. Educ.*, vol., 21, no. 2, pp. 155–310.

McShane, P. and Anderson, B. (2002) *Beyond Establishment Economics: No Thank-you Mankiw* (Axial Publishing, Vancouver).

McShane, P. (2013) *Futurology Express* (Axial Publishing, Vancouver).

Mielczarek, E. V. (2006) Resource Letter PFBi-1: Physics frontiers in biology, in *Am. J. Phys.*, vol. 74, issue 5, May, pp. 375–381.

Morgenstern, A. P. and Velásquez, N. (2013) Resource Letter MPCVW-1: Modeling Political Conflict, Violence and Wars: A survey, in *Am. J. Phys.*, vol. 81, issue 11, Nov., pp. 805–814.

O'Shea, P. (2012) Physics and Medicine 4, Future Medicine shaped by an interdisciplinary new biology, in *The Lancet*, vol. 379, April 21, pp. 1544–1550.

Pearsona, A. J., Pizzutoa, J. E., and Vargasb, R. (2016) Influence of run of river dams on floodplain sediments and carbon dynamics, *Geoderma*, vol. 272, 15, June, pp. 5163.

Rief, R. (1999) Thermal physics in the introductory physics course: Why and how to teach it from a unified atomic perspective, *Am. J. Phys.*, vol. 67, issue 121, Dec., pp. 1051–1062.

Riendeau, D. (2014) The languages of physics education, *Phys. Teach.*, 52, April, pp. 251–252.

Rossing, T. D. (1975) Resource letter MA-1: Musical acoustics, in *Am. J. Phys.*, vol. 43, issue 11, Nov., pp. 944–953.

Shute, M., (2010a) *Lonergan's Discovery of the Science of Economics* (University of Toronto Press, Toronto).

Shute, M. (2010b) *Lonergan's Early Economic Research* (University of Toronto Press, Toronto).

Stanleya, H. E., Amarala, L. A. N., Gabaixb, X., Gopikrishnana, P. and Pleroua V. (2001), Similarities and differences between physics and economics, *Physica A*, 299, pp. 1–15.

Stefan, V. ed. (1998) *Physics and Society, Essays in Honor of Victor Frederick Weisskopf by the International Community of Physicists* (Springer Verlag, New York).

Szpiro, G. G. (2011) *Pricing the Future: Finance, Physics, and the 300-Year Journey to the Black-Scholes Equation; A Story of Genius and Discovery* (Basic Books, New York).

Wang, H, Yang, Z., Saito, Y., Liu, J. P., Sun, X., (2006) Inter-annual and seasonal variation of the Huanghe (Yellow River) water discharge over the past 50 years: Connections to impacts from ENSO events and dams, *Glob. Planet. Chang.*, April, vol. 50, issues 3–4, pp. 212–225.

Weisskopf, V. (2003) The Privilege of Being a Physicist, in *Phys. Today, Am. Inst. Phys.*, Feb., pp. 48–52.

Zarfl, C., Lumsdon A. E., Berlekamp, J., Tydecks, L., and Tockner, K. (2015) A global boom in hydropower dam construction, *Aquat. Sci.*, Jan., vol. 77, issue 1, pp. 161–170.

Chapter 9

Implementation

9.1 Introduction

This chapter provides diagrams which highlight various aspects of functional collaboration. Much as biochemistry needs constellations of diagrams, developments toward, and in, the new standard model also will need groupings of diagrams – for various cross-sections, structurings, sub-structurings, sub-structurings of sub-structurings, and so on[1]. This chapter has only a few diagrams, some of which are McShane's metagrams, $W1$, $W2$, ...[2] Some of the diagrams presented are obtained by combining different elements from some of the basis metagrams $W1$, $W2$, ... A few of the diagrams here, though, are new – elementary but, I think, helpful. Note also that, in order to fit the ordering of topics presented in this book[3],

---

[1] Diagrams are normal in the sciences. There is also Lonergan's advice to his students: "But the aim of discursive reasoning is to understand; and it arrives at understanding not only be grasping how each conclusion follows from premises, but also by comprehending in a unified whole all the conclusions intelligibly contained in those very principles. Now this comprehension of everything in a unified whole can be either formal or virtual. It is virtual when one is habitually able to answer readily and without difficulty, or at least 'without tears,' a whole series of questions right to the last 'why?' Formal comprehension, however, cannot take place without a construct of some sort. In this life we are able to understand something only by turning to phantasm; but in the large and more complex questions it is impossible to have a suitable phantasm unless the imagination is aided by some sort of diagram. Thus, if we want to have a comprehensive grasp of everything in a unified whole, we shall have to construct a diagram in which are symbolically represented all the various elements of the question along with all the connections between them" (Lonergan [2002], p. 151).

[2] McShane [2008]. McShane's diagrams $W1$, $W2$,... are fundamental.

[3] The book rises on (preliminary) description of eight distinct tasks.

I introduce a different labeling: *D1*, *D2*, and so on[4]. Diagrams *D1*, *D2* and *D3* are different renderings of the main functional cycling; *D4a* and *D4b* are for economics, a sub-structuring that, as it happens, also was discovered by Lonergan; *D5a* and *D5b* point to what eventually will be a functionally-structured present state (statistically speaking); *D6a*, *D6b* and *D6c* are for statistical states in history, present and transitional; and *D7* represents a future geo-historical stochastic sheaf structuring.

9.2 Central Structurings

D1: This is a simplified rendering of the eight functional specialties. It also indicates the attentiveness of functional research to events and occurrences of the world, as well as the mediation and outreach of functional communications.

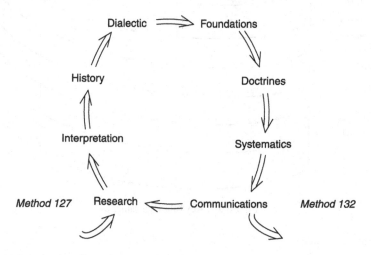

Fig. 9.1. Functional Specialties. "Functional specialization distinguishes and separates successive stages in the process from data to results" (Lonergan [1975], p. 126). The division of labor is to be "a normative pattern of recurrent and related operations yielding cumulative and progressive results" (Lonergan [1975], pp. 4, 5).

---

[4] The ordering of McShane's diagrams is based in a more fundamental rationale. See note 2.

D2: Front-line scholars will have an increasingly adequate heuristics of aggreformic entities[5]. This will be integral to the entire functional structuring, and can be represented by layerings in diagram *D1*:

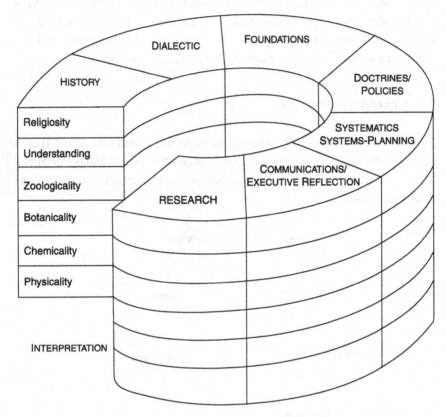

Figure 9.2. Keyhole Diagram[6].

---

[5] This is a major scientific challenge of our time. See Lonergan [1992], ch. 8 and the transposition to sec. 15.3. The issue is discussed in many works by McShane – see, e.g., McShane [1998], pp. 120–123. See also Quinn, *Invitation to Generalized Empirical Method* [2017]. There is physicality; chemicality; botanicality; zoologicality; understanding; (and religiosity). Religiosity, a core orientation, is, however, not treated in this book.

[6] I have taken the liberty of rotating the Keyhole diagram (McShane [1998], p. 110) so that, as in other diagrams in this book, functional dialectics and functional foundations are at the top; and the central functional cycling is clockwise.

D3: The next diagram is adapted from McShane's metagram *W5*[7]. I would not presume to know McShane's rationale for using *H*'s. For the present introductory context, *F1* stands for functional research, *F2* for functional interpretation, and so on. That is, each *Fi* stands for the *i*[th] functional specialty. The "lanes" are a relatively permanent complexification of diagram *D1*. For transition toward the new standard model will be gradual; and even when that occurs, there will degrees of competence.[8]

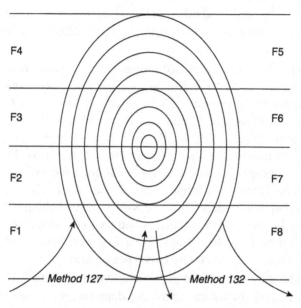

Figure 9.3. *W5* with *F* notation. "There are various ways of reading this diagram, but the main point is to think at least of a single 'track' that links the functional specialties, each specialty passing the baton to the next in the 'race' for progress. The outer track can be considered to be the best up-to-date model of the collaboration that is to emerge, perhaps in the next century. I think it useful to associate the set of tracks with the light spectrum: the first track is 'red,' a rough-neck effort to collaborate that will characterize our early efforts: the outer track is to have the sophistication of 'indigo.'"[9]

---

[7] See note 2. See also McShane [2007], fig. 1, p. 204. This figure was used in the design of the book cover.

[8] Readers may wonder why there is a recurring pattern of "four levels" in diagrams for functional specialization. This is a basic and also non-trivial issue. See Epilogue, Section E.2, Control of Meaning. See also, Lonergan [1975], sec. 5.3, Grounds of the Division, pp. 133–136.

[9] McShane, [2008], *W5*, pp. 9–10.

The laneways of *D3*, of course, do not represent strict delineations. We can, though, anticipate genera and species of growth across three main differentiations: description; theory; and control of meaning of *tentative universal viewpoints*[10]. There will, for example, be engineers who have some descriptive grasp of the standard model, but who focus on engineering projects in industry of the day. But, there will also be front-line scholars who have a special interest in engineering science and who contribute, luminously, to the new standard model. And, there will be middle cases. All such work will contribute to the World Caring Functional Enterprise.

D4a* and *D4b: The Science of Economics. As is evident from Chapter 8, progress partly depends on and includes economics. Economics will be a sub-structuring of functional collaboration. As noted in Chapter 8[11], present-day Establishment Economics is contributing to ongoing global economic and cultural crises. Descriptions such as Mankiw's Ten Principles[12] neither explain nor resolve current problems. More than 70 years ago, Lonergan discovered an economic theory[13] verifiable in businesses and economies, large, medium and small. The present global economic mess partly depends on errors in Establishment Economics, identified by Lonergan. For instance, contrary to the usual hypothesis that there are "Firms and Households,"[14] a key verifiable (and ultimately explanatory) result is: "There are two types of firm."[15]

The reader may not wish to pursue these matters here[16]. However, for completeness, and for future reference, diagrams are given for the new science of economics discovered by Lonergan. A first is *D4a*.

---

[10] Regarding 'universal viewpoint, see Lonergan [1992], 587–591. The expression 'tentative universal viewpoint' was introduced by McShane and is discussed in several of his works. See, e.g., McShane [2004], p. 2.

[11] See note 25 of Chapter 8.

[12] See, e.g., Mankiw [2015], Chapter 1.

[13] Lonergan [1998], [1999]; and Shute [2010].

[14] See, e.g., Mankiw [2015], Figure 1, p. 23. Inclusion of other factors such as government and international trade complexifies, but does not remove initial oversights.

[15] McShane [2014], ch. 1, The Key Issue, pp. 5–14. See note 25 of Chapter 8 for additional references.

[16] McShane has various introductions to Lonergan's economics. See McShane [2014], references therein, and references in note 25 of Chapter 8.

D4a:

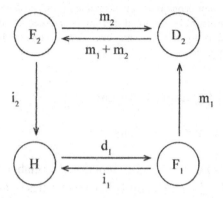

Figure 9.4a. There are two types of firm[17].

In *D4a*, the two types of firm are represented by F_1 and F_2. Diagram *D4b* (more complex) is for Lonergan's complete surplus-basic field equations and includes a normative central redistributive zone.

D4b:

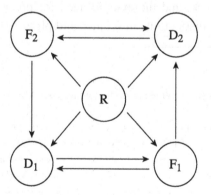

Figure 9.4b. Diagram of Transfers between Monetary Functions[18].

---

[17] McShane [2014], pp. 13–14.

[18] Keeping the notation of *D4a*, diagram *D4b* is a simplified version of the diagram in Lonergan [1998], Transfers between Monetary Functions, p. 258. The symbol '*R*' stands for *redistributive function*. See Lonergan [1998]: Part 1, ch. 5, sec. 34, pp. 62–68, diagram p. 64; and Part 3, ch. 16, sec. 8, pp. 252-258, diagram p. 258.

Diagram *D4a* provides a basis for

the beginnings of a new economics of measurable flows, one that would yield norms of financing, of profit in both normal and innovative economies.[19] ... (But,) the fundamental need is to sort out this beginning, before adding banks, taxes, international trading, etc. etc.[20]

9.3 Functionally Structured Present

D5a: Convergences and Expansions

As pointed to in *D2*, in mature functional collaboration, generalized empirical method will yield an up-to-date heuristics of aggreformism. It is not that differences in sciences will vanish. However, inquiry will be under a control of meaning that will include differentiation of eight fundamentally distinct tasks.

Diagram *D5a* points to the fact that, in the first phase, transitions will involve convergences; and in the second phase, expansions[21]. Diagram *D5a* also points to functional specialization resolving various contemporary issues, including: questions about the nature and scope of interdisciplinary work; and the evident need for precision in identifying commonalities and distinctions in science and philosophy of science.[22]

---

[19] See note 17.

[20] See note 17.

[21] For additional details on convergences and "slopings," see McShane [2008], *W6*, pp. 10–11.

[22] At this stage of history, interdisciplinary efforts tend to be (semi-) random. (See, e.g., statistical methods presented in Murray, Ke and Börner [2006]; and statistical results in Bishop et al., [2014]. We can expect distributions of future collaborations to gradually take on a functional ordering, C_{ij}. See Functional Communications Markov Process, Figure 8.2. See also Section 9.4, on stochastics.) Philosophical reflection on interdisciplinarity remains inconclusive. (Comparing present-day views of interdisciplinary science with actual (interdisciplinary) performance, "(i)t is quite clear that typical contributions to the unity of science do not answer the currently most pressing questions about interdisciplinarity" (Grüne-Yanoff and Mäki [2014], p. 53). Contemporary philosophy of science lacks consensus and precision about science and philosophy of science. ("(I)t is obvious, that by our lights, there is no sharp line between philosophy of physics and physics itself" (Butterfield and Earman [2007], p. xviii). "There is no clear boundary between philosophy

By way of illustration, an astronomical laboratory may detect surprising features in Balmer radiation distributions. But, telescopes, apparatus and astronomical objects are physical and chemical matter and, in order to identify astronomical matter, one needs both chemistry and physics. Note, too, that for sufficiently similar data, one does not need a new interpretation.

Where functional interpretation looks to the meanings of individuals, functional history will bring interpretations together. That is, it will attempt to explanatorily identify (genetic-dialectic) sequences that were actually taken up and influential in the community[23].

But, there are different histories written by different historians with different viewpoints. And such histories can focus on different aspects, of, e.g., experimental physics, theoretical physics, philosophy of physics, physics and art, physics and society, and so on.[24] And so, we get a glimpse of a further gathering of results that will be called for in the task called functional dialectics[25].

The example above supposed Balmer-line data communicated to the community. Potentially significant data, though, also can be found in, for example, the writings of philosophical reflection. And explanatory interpretation of such work calls upon all aggreformic layerings[26].

---

of biology and theoretical biology" (Garvey [2007], p. xii). "(S)ome philosophical work shades off into science; there is no sharp border between them" (Godfrey-Smith [2014], p. 1).) See also notes 12 and 13 of Chapter 4. These observations describe commonalities, but also beg the question. All of philosophy of science (and all of philosophy) will enjoy the normative advantages of an eightfold division of labor. Within the new ethos, all of philosophy will be transformed, and in particular will be essentially free of present-day conceptualist tendencies. Instead, it will be fundamentally empirical, in the sense that it will have control of meaning normal in generalized empirical method.

[23] Chapter 3.

[24] Once the new standard model for collaboration is operative, there will be histories about workings in the standard model: histories about the common plane of meaning, histories about functional research; histories about functional interpretations; and so on.

[25] Chapter 4; and Lonergan [1975], p. 250.

[26] Chapter 2.

So, in the transitioning from research to interpretation; interpretation to history; and history to dialectics, we find a general convergence. In the transition from research to interpretation, a multiplicity of data is subsumed by a single interpretation. In the transition from interpretation to history, many interpretations are brought together in genetic-dialectic sequences. In the transition from history to dialectics, one seeks to attain a single up-to-date basic position and intelligent view of all, a best-story and best-basic-directions.

Note also that in the transitioning from research through dialectics, it becomes increasingly necessary to appeal to the full heuristics of aggreformism.

What about the four future-oriented functional specialties, functional foundations, doctrines, systematics and communications? Instead of convergence, communications C_{56}, C_{67}, C_{78} and C_{89} will involve successive expansions.

Functional foundations will provide "general solutions," best fundamental directions not tied to any time[27]. Doctrinal development will take up "general solutions," and work out (historically grounded) doctrinal orderings. Functional doctrines will provide pragmatic truths about progress and development[28]. Genetic systematics will draw correctly and contra-factually on discoveries and strategies of the past, to envisage ranges of time-ordered possibilities[29]. Doctrines will find explanatory significance in genetic systematics. Functional communications will involve still further expansions. For, taking advantage of up-to-date genetic systematics, there will be creative solutions needed and communicated to meet vast ranges of local and global needs.

---

[27] McShane [1998], p. 106.
[28] McShane [1998], p. 106.
[29] McShane [1998], p. 106.

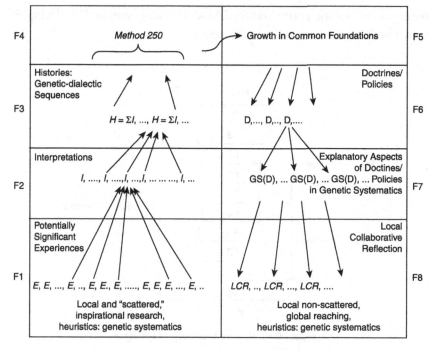

Figure 9.5a. Convergences and Expansions[30].

D5b: Raising Standards of Living

As described in *D1*, the first phase (research, interpretation, history and dialectics) will be past-oriented, and the second phase (foundation, doctrines, systematics and communications) will be future-oriented. Diagram *D5b* is: partly inspired by the Stair Way diagram (Figure 8.1); points to convergences and expansions of *D5a*; and includes the time component of Figure 8.2. Both phases will be part of a present enlightenment and enliftment cone-zone, laboring to raise standards of

---

[30] See also Lonergan [1975], p. 142: "As the first phase rises from the almost endless multiplicity of data first to an interpretative, then to a narrative, and then to a dialectical unity, the second phase descends from the unity of a grounding horizon towards the almost endlessly varied sensibilities, mentalities, interests and tastes of mankind. The descent is, not properly a deduction, but rather a succession of transpositions to ever more determinate contexts."

living for all, for the entire Standard Model (where 'standards of living' is meant both as measure and as achievement).

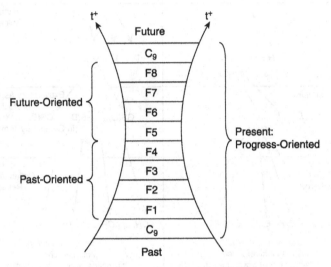

Figure 9.5b. Raising Standards of Living.

9.4   Gradual Emergence of Functionally Coherent States

As already discussed in this book, pressures of history gradually have been revealing the presence eight main tasks. In physics, for example, there is ongoing collaboration between Experimental Physics and Theoretical Physics, pre-cursor of a future functional research and functional interpretation. As discussed in earlier chapters, evidence for the other tasks can be found in, for example, historical work; philosophy of physics; groups attempting to meet local needs of society; and so on.

Each specialty will have its own emerging aims and strategies. In future, we can expect that making front-line contributions to any one task will be as much as can be expected from an individual investigator. Does this suggestion seem surprising? Even now, serious contributions to, say, experimental physics usually do not also include theoretical developments as such, and *vice versa*. O'Raifeartaigh's work on the dawning of gauge theory does not also attempt new theoretical development; and nor does it provide new experimental data. And, in practice, questions about, say, present local-global needs usually take histories (written and lived)[31] as background.

While four centuries or so of progress in physics has involved a developing collaboration between experimental and theoretical physics[32], and other tasks as well are coming into view, this has not yet involved an across-the-board control of meaning. This is especially noticeable in areas such as hermeneutics, historical studies, philosophy of science, and generally, the human sciences. And so it is that, at this time in history, it is normal for two or more fundamentally distinct tasks to be inadvertently present in individual works, in *ad hoc* ways. This is not merely a lack of control of meaning, but undermines the potential effectiveness of such scholarship.[33]

Diagram *D6a* provides an image of what this looks like in two present-day individual works, relative to the future standard model. Diagram *D6b* represents a typical present-day distribution of such multi-tasking within a sample space of the literature[34]. Diagram *D6c* partly was inspired by Figure 10.1 of Ian Lawrie's book, and by other diagrams of vortices, pattern formation and equilibrium states in physics and biology[35]. But, of course, the diagram is not about such aggregates, but is about history. Diagram *D6c* represents a time-ordered sequence of typical distributions, a gradual transitioning (over centuries or perhaps millennia)[36] out of the

---

[31] See Chapter 8.

[32] Leading advocates for string theory question the need of normative collaboration between theoretical and experimental physics. My comment, however, is non-controversial. I simply point to what has been standard practice since Galileo's discovery of the law of falling bodies. This is seen, for example, in ongoing international collaborations that appeal to data obtained from CERN and other particle accelerators, as well from (hundreds of) astronomical observatories (space-based, airborne, ground-based and underground-based). See also Introduction.

[33] For brief discussion of two examples, see Appendix A.

[34] Ultimately, we also need to include locations and times. See Diagram *D7*.

[35] Lawrie [1990], fig. 10.1, p. 219.

[36] See note 10 of Chapter 7.

present state (statistical) of confusion (Axial Times[37]) toward a (third) stage[38] of history wherein functional coherence will be normalized (statistical) in the Academy and world communities. Diagram *D6c*, then, also represents an increasing dominance of modes (statistical) that have functional coherence. It may require several millennia to reach mature functional collaboration. However, advantages of eightfold collaboration can be expected to become increasingly evident within a few decades of initial efforts.

D6a:

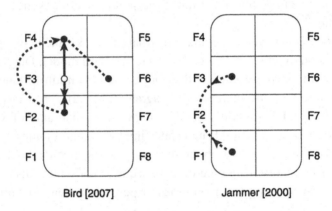

Bird [2007]   Jammer [2000]

Figure 9.6a. Mixing of tasks evident in two works[39].
O : implicit. non-directed
● : explicit, non-directed
− − − : implicit. linkage involved
——— : explicit, linkage involved.
→ : directed

---

[37] McShane [1998], pp. 1–2. See also, Preface.
[38] See Lonergan [1975], Sections 3.10, 3.11, pp. 85–99; and notes 19 and 20 in Preface.
[39] The symbolism is not to indicate strict delineations, but some dominant modes in these particular works. See Appendix A, regarding the book, Jammer [2000] and the article, Bird [2007].

D6b: Typical distribution of non-functional works

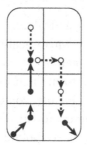

Figure 9.6b. Typical distribution of non-functional works[40].

O : implicit. non-directed

● : explicit, non-directed

– – – : implicit. linkage involved

——— : explicit, linkage involved.

→ : directed

D6c: Decreasing scattering and turbulence; increasing functional coherence.

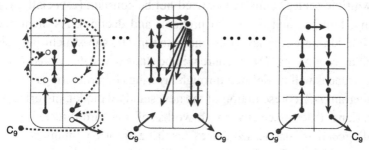

| Present: Meanings mixed and scattered | Preliminary decades and centuries of making efforts to work within eight-fold division of labor: functional modes gradually emerging. | Future millennia: Dominant etbos functional, functional modes, e.g. $C_{i,i+1}$, relatively stable, yielding cumulative and progressive resuls. |

Figure 9.6c. Decreasing turbulence and increasing functional coherence[41].

O : implicit. non-directed

● : explicit, non-directed

– – – : implicit. linkage involved

——— : explicit, linkage involved.

→ : directed

---

[40] Much as in *D6a*, the symbolism is not to indicate strict delineations, but some dominant modes. Here, though, in *D6b*, modes are from representative samples of subsets in the literature.

[41] Once functional collaboration stabilizes, all communications C_{ij} will be luminously operative. In diagram *D6c*, only the central flows (C_{91}), C_{12}, ..., C_{78}, (C_{89}) are in the figure.

9.5   Stochastic Sheaf-Theoretical Geo-History

D7: We live and work at times and places, in communities, in history. Diagram $D7$ represents the geo-historical-sheaf structure, with local structurings. More refined notation eventually will be needed. A refinement would be to represent time intervals in a more usual way, for example, by ΔT. For this introductory context, I keep symbolism to a minimum, and so represent a time interval simply by T. In the same way, X represents a spatial interval, where both X and T are descriptive. The world at time interval T_a is represented by a sphere, with locations X (department, institute, town, region, etc., as the case requires). A larger sphere is for the world at a later time interval T_b, $a \leq b$. Local structurings will be determined by functional specialties F_i, $i = 1, ..., 8$, communications C_{ij}, $i, j = 1, ..., 9$, and sub-structurings such as those indicated in previous sections.

Sub-structurings, sub-structurings of sub-structurings, and so on, have their own time-scales. Time-scales need not be constant (especially, e.g., in botany, biology, and in human progress and decline). As mentioned above, in addition to the layerings indicated in $D1$, sub-structurings also will include economics ($D4b$). There will be statistical states whose events and occurrences will be defined through layerings of structurings.

In diagram $D7$ representation of the new standard model, histories will be foliations determined by mesh-works of outward-directed (time-directed) genetic-dialectic sequences resulting from communications of the form $C_{ij}(X_k, X_l)(T_a, T_b)$, between individuals at location intervals X_k and X_l, in time intervals T_a and T_b. When times are different, communication is one-way, such as when reading the work of a scholar from an earlier time in history. A cross-section for time interval T is given by communications of the form $C_{ij}(X_k, X_l)(T) = C_{ij}(X_k, X_l)(T, T)$, for time interval T. A cross-section for location X is given by communications of the form $C_{ij}(X)(T_a, T_b) = C_{ij}(X, X)(T_a, T_b)$, communications in location X over time intervals T_a and T_b. In particular, economic activity ($D4b$[42]) will be conceived, affirmed and implemented[43] locally and globally.

---

[42] See note 18.
[43] Lonergan [1992], p. 416.

With the gradual emergence of the new standard model of global collaboration, world populations increasingly will embody the ethos of an eightfold division of labor in which, in particular, elite scholars will be relatively luminous in their presence and caring for the peoples of the world.

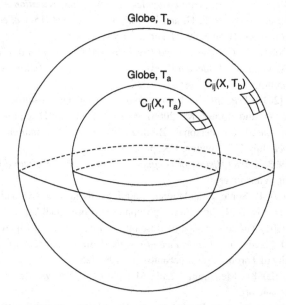

Figure 9.7. Geo-historical stochastic sheaf with local structurings.

9.6 References

Anderson, B. (2004) Economics as if Local Community Mattered, *Catholic Rural Life*, Spring, pp. 16–19.

Bird, A. (2007) What is Scientific Progress? *Noûs*, vol. 41, issue 1, pp. 92–117.

Bishop, P. R. Huck S. W., Ownley, B. H., Richards, J.K and Skolits, G.J. (2014) Impacts of an interdisciplinary research center on participant publication and collaboration patterns: A case study of the National Institute for Mathematical and Biological Synthesis, *Res. Eval.* 23, pp. 327–340.

Butterfield, J. and Earman, J., eds. (2007) *Philosophy of Physics, Part A* (Elsevier, Amsterdam).

Garvey, B. (2007) *Philosophy of Biology* (Acumen, Stocksfield Hall).

Godfrey-Smith, P. (2014) *Philosophy of Biology* (Princeton University Press, Princeton).

Grüne-Yanoff, T. and Mäki, U. (2014) Introduction: Interdisciplinary Model Exchanges, *Studies Hist. Phil. Sc.*, 48, pp. 52–59.

Jammer, M. (2000) *Concepts of Mass in Contemporary Physics and Philosophy* (Princeton University Press, Princeton, NJ).

Lonergan, B. (1975) *Method in Theology* (Darton, Longman & Todd, London).

Lonergan, B. (1998) *For a New Political Economy*, McShane, P. ed., vol. 21 in *Collected Works of Bernard Lonergan* (University of Toronto Press).

Lonergan, B. (1999) *Macroeconomic Dynamics: An Essay in Circulation Analysis*, ed. by Lawrence, F. G., Byrne, P. H. and Hefling, Jr., C. C., vol. 15, *Collected Works of Bernard Lonergan* (University of Toronto Press, Toronto).

Lonergan, B. (2002) *The Ontological and Psychological Constitution of Christ*, vol. 7 in Shields, M. G., trans., *Collected Works of Bernard Lonergan* (University of Toronto Press, Toronto).

Mankiw, N. G. (2015) *Principles of Macroeconomics* (Cengage, Boston).

Murray, C., Ke, W. and Börner, K. (2006) Mapping Scientific Disciplines and Author Expertise Based on Personal Bibliography Files, *Information Visualization Conference*, July 5–7, London, UK, pp. 258–263.

McShane, P. (1998) *A Brief History of Tongue, From Big Bang to Coloured Wholes* (Axial Publishing, Vancouver).

McShane, P. (2004) Some Foundational Pointings Regarding Evaluation, no. 9 in *Quodlibets*, nos. 1–21, http://www.philipmcshane.org/quodlibets/.

McShane, P. (2007) The Importance of Rescuing *Insight*, art. 12 in Liptay, J. L. Jr. and Liptay, D. S., eds., *The Importance of Insight, Essays in Honour of Michael Vertin* (University of Toronto Press, Toronto), pp. 199–225.

McShane, P. (2008) Metagrams and Metaphysics, *Prehumous 2*, http://www.philipmcshane.org/.

McShane, P. (2014) *Piketty's Plight and the Global Future, Economics for Dummies* (Axial Publishing, Vancouver).

Nedel, A. (2013) *Husserl, Cantor & Hilbert: La Grande Crise des Fondements Mathématiques du XIXeme Siecle* (Cornell University Library, arXiv:1311.1524 [math.HO]).

Quinn, T. (2016) Interpreting Lonergan's 5th Chapter of *Insight*, ch. 2 in Brown, P. and Duffy, J. eds., *Seeding Global Collaboration* (Axial Publishing, Vancouver), pp. 29–44.

Quinn, T. (2017) *Invitation to Generalized Empirical Method* (World Scientific Publishing, Singapore).

Shute, M. (2010) *Lonergan's Discovery of the Science of Economics* (University of Toronto Press, Toronto).

Epilogue

Choosing our Future

E.1 Getting Our Bearings

Where do we go from here? One option is to ignore Lonergan's discovery.[1] However, given the abundance of data already adverted to in a range of areas[2], it is becoming increasingly probable that scholars in established areas will begin to notice that Lonergan was on to something important. Indeed, as becomes increasingly evident and self-evident in the book, the division of labor envisioned will be omni-disciplinary.

A different option, then, is to take the discovery to heart. Help can be had from works just cited[3] and, hopefully, from this book as well. But, the functional division of labor is not yet operative. And, as quoted also in the Introduction, there is Aristotle's observation[4]. We will learn by doing.

---

[1] Unfortunately, so far, that has been the choice of many scholars in Lonergan Studies. See Brown and Duffy [2016], note 75, p. xx. Fundamentally, the omission would be comparable to admirers of Einstein's early work ignoring his later breakthrough to General Relativity. (See Preface, second last paragraph.) Although, as recounted in An Unpopular Theory (Eisenstaedt [2006], ch. 11, pp. 244–254), general relativity also was not easily accepted. And, Lonergan's discovery was methodological. Transitioning to functional collaboration will take time. Still, it has been almost 50 years since Lonergan's discovery was first published in Lonergan [1969], and so it is hoped that before too much longer, Lonergan Studies will turn attention to Lonergan's discovery. See Preface, note 25.

[2] This includes works cited in note 2 in the Preface.

[3] See note 2.

[4] "For the things we have to learn before we can do them, we learn by doing them" (Introduction, note 26; Bartlett and Collins [2011], Book 2, p. 26; or Ross [2001], Book II, par. 1).

To help toward beginnings in the new methodology, a recent book, *Seeding Global Collaboration*[5], is a collection of articles where each author tried their hand at keeping to a functional focus. Some details on that multi-author project are described in the Editors' Introduction to that book, a helpful article regarding the task before us, the transition to effective collaboration[6].

Of course, we cannot expect immediate efficacy. However, as the papers in *Seeding Global Collaboration* bring out, even initial efforts in the new methodology significantly increase pressure for development in control of meaning well beyond present norms, including beyond the important but only initial achievement of being able to describe eight main tasks present in the Academy.

E.2   Control of Meaning

What becomes increasingly (self-) evident is that part of what is needed is that we read and "struggle with some such book as *Insight.*"[7] By "read," I mean read and struggle in the way that a science student reads and struggles with an advanced textbook (or for that matter, the way a musician struggles to slowly grow toward mastery of a book of *Études*[8]). In *Insight*, the author himself draws attention to the needed strategy[9]. The reader is asked to work through series of graduated exercises in self-attention, through ranges of contexts that begin with, and build on, classical and modern sciences, and more.

This type of book-structuring will be familiar to readers in physics, mathematics and other sciences. To read any advanced textbook (in physics, for example), one needs to struggle through the exercises. Certainly one does not reach initial insights in physics by first reading Lawrie's well-known book, *A Unified Grand Tour of Theoretical Physics.*[10] The book *Insight* also is *A Unified Grand Tour ... of Human Understanding*. Lonergan's *Insight*, though, was the achievement of a genius-lone climb. Like any graduate text, the book needs a massive

---

[5] Brown and Duffy [2016].

[6] Brown and Duffy [2016], Editors' Introduction, pp. i–xxii.

[7] Lonergan [1992], note 2, p. 7.

[8] These are just some examples. Look also, for example, to development in one of the arts, sciences or sports.

[9] Lonergan [1992], last paragraph of ch. 1, pp. 55–56.

[10] Lawrie [1990].

underlay of more elementary books and pedagogy to help us prepare for and gradually climb toward a reaching control of meaning densely described in steeply ascending chapters[11].

While the needed prior books are not yet present, we can begin with strategic forays. There is, for instance, the discovery of Archimedes' principle, which, in fact, is the first example that begins the first paragraph of the first section of *Insight*.[12] But, we needn't struggle alone. There are pedagogical works already available to help us make beginnings in self-attention (in science, mathematics, and other areas)[13].

Even though our results are, for now, only descriptive, by attending to one's experience one can discover an invariant ordering of elements in the dynamics of human knowing and doing. Two key diagrams for this are in Appendix A of *Phenomenology and Logic*[14], simplified versions of which are as follows:

Dynamics of Knowing

State of Wonder ⇑	Is?	! Reflective insight	Judgement (Yes/No)
State of Wonder ⇑	What?	! Direct insight	Inner formulation
State of Wonder	Sense		

Dynamics of Doing

State of Wonder ⇑	Is-to-do?	! Reflective insight	Judgement of Value
State of Wonder ⇑	What-to-do?	! Direct insight	Inner formulation
State of Wonder	Sense, images, known facts		

Figure E.1. Dynamics of Knowing and Doing.[15]

---

[11] Much of McShane's published work is devoted to such efforts. But, see, in particular, McShane [1970] and McShane [1998].

[12] Lonergan [1992], pp. 27ff.

[13] McShane [1975]; McShane [1998]; and Benton, Drage and McShane [2005].

[14] Lonergan [2001], Appendix A, pp. 322–323.

[15] See also, McShane [1975], pp. 15, 48; and Benton, Drage and McShane [2005], pp. 64, 74; and Benton [2008], p. 62.

Much as we can describe eight main tasks in the Academy, here too, the result is empirical. One can describe and name four main question-forms, and so five ways in human knowing:

(1) Experience; (2) What?; (3) Is?; (4) What-to-do?; (5) Is-to-do?

Figure E.2. Five ways in human knowing[16]

There are *five* ways. How, then, are there *eight* functional specialties? Within each functional specialty there will be all five ways of human knowing. The shift to functional specialization, though, will be the emergence of a new ordering, a new differentiation in the Academy wherein each specialty will have its own functionally distinct task. And, through elementary exercises in self-attention, one can make beginnings toward identifying "a correspondence between the operations of the human subject and its intended objects in each phase"[17] of the functional division of labor.

The correspondence can be expressed in the following diagram, where the letters are for the eight functional specialties and the numbers are for five levels of human knowing, respectively[18]:

R	I	H	D	F	D	S	C
1	2	3	5	5	3	4	1

Figure E.3 Correspondence between functional specialties and operations of human subject.[19]

These few paragraphs are not pedagogical. I am pointing to extensive and creative learning needed, beginnings of which are possible for us with

---

[16] This is an abbreviated map. "Grasping the full meaning of those names and how they interlock …is a lifelong climb" (Benton, Drage and McShane [2005], p. 95).

[17] Benton [2008], p. 105.

[18] The 'D' in the first phase is for *Dialectics*, and in the second phase is for *Doctrines*.

[19] McShane [1998], p. 100; Benton [2008], p. 106

the help of some of the books already cited.[20] Eventually, explanatory results will emerge within a maturing generalized empirical method.

E.3 Our Choice to Make

This Epilogue started with a question: Where do we go from here? Do we continue with the *status quo*, in patterns of semi-random collaboration that do not distinguish what in fact are distinct tasks; that do not cultivate control of meaning; that allow for more or less endless debate about fundamental issues; that contribute to cumulative decline in global cultures? Or, do we make efforts toward implementing what pressures of history already are revealing in major areas of inquiry, namely, the possibility of a normative division of labor around what evidently and self-evidently are eight distinct tasks; the possibility of a basis of collaboration and global care that promises to be statistically effective; a collaboration that is oriented "to make operative the timely and fruitful ideas that otherwise are inoperative,"[21] that is oriented to "protect the future,"[22] a collaboration that is structured to promote progress[23] and offset decline,[24] and, indeed, a collaboration that, in community, will increasingly manifest our very human subjectivity? The choice is ours to make.

E.4 References

Bartlett, R.C. and Collins, S. C., trans. (2011) *Aristotle's Nichomachean Ethics. A New Translation* (University of Chicago Press, Chicago).

Benton, J. Drage, A. and McShane, P. (2005) *Introducing Critical Thinking* (Axial Publishing, 1st prtg, Judique, Nova Scotia, subseq., Vancouver).

Benton, J. (2008) *Shaping the Future of Language Studies* (Axial Publishing, 1st printing, South Brookfield, Nova Scotia, subseq., Amazon).

Brown, P. and Duffy, J. (2016) *Seeding Global Collaboration* (Axial Publishing, Vancouver).

---

[20] See note 13. Regarding creative learning, see also ch. 5, note 31.

[21] Lonergan [1992], p. 264.

[22] Lonergan [1992], p. 265.

[23] Lonergan [1972], p. 44.

[24] Lonergan [1972], p. 44.

Lawrie, I. D. (1990) *A Unified Grand Tour of Theoretical Physics* (IOP Publishing, Bristol).

Lonergan, B. (1969) Functional Specialties in Theology, *Gregorianum*, vol. 50, pp. 485–505.

Lonergan, B. (1992) *Insight: A Study of Human Understanding*, vol. 3 in Crowe, F. E. and Doran, R. M. eds. *Collected Works of Bernard Lonergan* (University of Toronto Press, Toronto).

Lonergan, B. (2001) *Phenomenology and Logic: The Boston Lectures on Mathematical Logic and Existentialism*, ed., Philip J. McShane, vol. 18 in *The Collected Works of Bernard Lonergan* (University of Toronto Press, Toronto).

McShane, P. (1970) *Randomness, Statistics and Emergence* (University of Notre Dame Press, Notre Dame).

McShane, P. (1975) *Wealth of Self and Wealth of Nations. Self-Axis of the Great Ascent* (Exposition Press, Hicksville, NY).

McShane, P. (1998) *A Brief History of Tongue. From Big Bang to Coloured Wholes* (Axial Publishing, 1st prtg, Halifax; subseq., Vancouver).

Quinn, T. (2017) *Invitation to Generalized Empirical Method* (World Scientific Publishing, Singapore).

Ross, W. D. tr. (2001) *The Nicomachean Ethics* (Virginia Tech, Blacksburg).

Appendix A

Discerning (Pre-) Functional Content

A.1 Introduction

Techniques in contemporary philosophy of science include combinations of: logical arguments about *science, scientific knowledge, concepts in science, progress in science,* and so on (where such terms are defined nominally); and logical arguments about imagined processes and imagined landscapes of alternate universes. In particular, such methods tend neither to advert to nor distinguish distinct tasks verifiable in one's experience in science. Consequently, the present-day literature is dominated by works that inadvertently transition through two or more of what in fact are eight fundamentally different tasks – doing so within sentences, paragraphs, sections and chapters. In other words, fundamentally different tasks are touched on inadvertently and often in only fragmentary or incomplete fashion.

Without denying internal logic of works, there is a basic and obvious inefficiency in such

totalitarian ambitions[1].

Among other things, inadvertently moving about among several tasks within a single work undermines the possibility of a coherent buildup that otherwise might contribute to progress. Contrast this with, for example, familiar patterns of collaboration between experimental physics and theoretical physics (not yet functional, but implicitly revealing two of the

---

[1] Lonergan [1975], p. 137.

eight tasks[2]). It is true that, at this time in history, like all of the sciences, experimental and theoretical physics also struggle within a general lack in control of meaning. But, after four centuries of progress, the ongoing effectiveness of the collaboration between experimental and theoretical physics can hardly be denied.

The first eight chapters of this book were to help raise awareness of not only two, but of all eight distinct tasks present in the scientific community. In this Appendix, I continue that effort by looking to a book by Max Jammer[3] as well as an article by Alexander Bird[4]. The exercise is to discern tasks implicit in these individual works. The exercise helps bring out core confusion in (present-day) methods. By the same token, the exercise also provides additional evidence for the need and potential advantage of an eightfold division of labor.

In fact, I look to only two selections of text from these works. For, note that the purpose here is not to single out these authors as such. The problems discussed are historical and global; and methods in these works represent something of what at this time is standard in the field. As part of our initial learning about the eight distinct tasks, similar (but increasingly precise) exercises will be needed across representative samples of the works of Jammer and Bird (as well across representative samples of works in zones in the Academy[5]). This brief Appendix, then, is by no means intended as a serious interpretation of the selected texts from Jammer and Bird. My immediate purpose is just to give some preliminary indication of a larger problem in the contemporary literature.

As discussed in sections A.4.1 and A.4.2, the works of both Jammer and Bird may have much to offer in the history and philosophy of physics. Jammer's book, for example, provides extensive detail on the story of *mass* in physics in the final four decades of the twentieth century. Bird's article reveals methodological anomalies that, if explored, could be a help in our struggle out of Axial Times[6] toward a third stage of meaning. But,

---

[2] See Chapters 1 and 2, above.
[3] Jammer [2000]. Max Jammer (b. 1915–d. 2010).
[4] Bird [2007].
[5] This will need to include re-investigation of one's own pre-functional efforts. Preliminary efforts in various areas can be found in references given in Preface, note 2.
[6] See Preface, notes 20 and 21.

in order to effectively determine their potential contribution to progress, these works (and, similarly, the works of other leading scholars) will need to be re-cycled within an increasingly luminous eightfold division of labor.

A.2   Max Jammer: *Concepts of Mass in Contemporary Physics and Philosophy*[7]

Max Jammer's book, *Concepts of Mass in Contemporary Physics and Philosophy*[8], is masterful in its careful, detailed and comprehensive scholarship. What, though, are what we might call its (pre-) functional contents?

Certainly, the book primarily is past-oriented, with an emphasis on theories since 1960. This can be seen by reading, but is also explicitly indicated in the Preface of the book:

> The book intends to provide a comprehensive and self-contained study of the concept of mass as defined, employed and interpreted in contemporary theoretical and experimental physics and as critically examined in the modern philosophy of science. It studies in particular, how far, if at all, present day physics contributes to a more profound understanding of the concept of mass. ... (I)t deals mainly with developments that occurred after 1960.[9]

In order to be more a little more precise than merely saying 'past-oriented,' let us look next to a few samples, which I label by JA, JB, JC, JD and JE – for Jammer, Statement A; Jammer, Statement B; and so on.

> JA: Clearly, the validity or acceptability of "new" theoretical constructs cannot be proved by showing that, in the limit, they converge or reduce to their corresponding classical analogues unless it is also shown that they satisfy the theoretical principles for the validity for which they have been contrived. For the convergence, or reduction, to their classical analogues is a necessary but not a sufficient condition for their acceptability.

---

[7] Jammer [2000].
[8] Jammer [2000].
[9] Jammer [2000], p. vii.

In the present case, these principles are those of the conservation of energy and the conservation of energy.[10]

JB: Einstein thus saw the real task of his 1935 essay on the mass-energy relation as demonstrating the following:[11]

JC: All the derivations of the mass-energy relation discussed so far have dealt only with the inertial mass of a body. But, as we already know, and as will be explained soon in greater detail, there is a conceptual distinction between inertial and gravitational mass.[12]

JD: Let us now turn to the philosophical problem concerning the mass-energy relation, that is, to the question of what, precisely, is the conceptual meaning of the equation $E = mc^2$. As we shall see, at least two different interpretations have been proposed in the literature on this subject.[13]

JE: The problem of the meaning of $E = mc^2$ became the subject of lively discussions after the Second World War, that is, after the atomic bombardment of Hiroshima and Nagasaki had so tragically revealed the ominous significance of the mass-energy relation for the destiny of humanity.[14]

The challenge here is to make a beginning in detecting tasks implicit in these statements. Partly statement JA is a claim about how physicists *have been* determining the acceptability of new theoretical constructs. In that respect, the claim partly leans toward (pre-) functional dialectics. However, as Jammer's conclusions reveal, the claim also relates this to a possible relatively permanent truth, descriptive of all possible theoretical constructs. What is evident, then, is a simultaneous reaching into the sixth task, namely (pre-) functional doctrines.

For statement JB, look to the first phrase: "Einstein thus saw ..." The paragraph that follows is (a fragment of) hermeneutical work (eventually to be functional interpretation) purporting to summarize what Einstein understood to be Einstein's task.

---

[10] Jammer [2000], p. 84,

[11] Jammer [2000], p. 84.

[12] Jammer [2000], p. 85.

[13] Jammer [2000], p. 85.

[14] Jammer [2000], p. 86.

Statement JC refers to prior developments and, at least partly, is historical. Although, the text that immediately follows JC does not provide an account of historical developments. Instead, the text transitions to interpreting something of what Einstein wrote about energy and gravitational mass.

In statement JD, what is the "philosophical problem"? The subsequent (last four) pages of Chapter 4 of the book (pages 86–89) discuss something of the broad intent. Those four pages include statements and quotations from scholars and students in physics, social studies and political philosophy (from Russia, Eastern and Western Europe and the USA, in the period 1935–1987) regarding mass, energy and Einstein's equation $E = mc^2$. In that respect, the main focus of those four pages is historical. However, while some continuities in topics are highlighted, developments and meanings are not investigated.

Statement JE suggests a relatively permanent "ominous significance of the mass-energy relation for the destiny of humanity," vague hintings of, among other things, (pre-) functional doctrines.

A.3   Alexander Bird: *What is Scientific Progress?*[15]

Alexander Bird is well known for his publications in the philosophy of science. The article that I look to here is: *What is Scientific Progress?*[16] As in Section A.2, I label quotations simply by BA, BB, ... – for Bird, Statement A; Bird, Statement B; and so on.

BA: In this paper, I will be comparing three approaches to characterizing progress.[17]

BB: I shall argue for the epistemic approach, the simple-minded cumulative knowledge account of progress.[18]

BC: Popper, somewhat forlornly, hoped that his adherence to verisimilitude would differentiate him from his anti-realist contemporaries. More recently, Ilkka

---

[15] Bird [2007].
[16] See note 15.
[17] Bird [2007], p. 92.
[18] Bird [2007], p. 93.

Niiniluoto (1987; 1989) has sought to build a whole philosophy of critical scientific realism on the foundations of his own account of verisimilitude.[19]

BD: Imagine a scientific community that has formed its beliefs using some very weak or even irrational method M, such as astrology. But by fluke this sequence of beliefs is a sequence of true beliefs. These true beliefs are believed solely because they are generated by M and they do not have independent confirmation. Now imagine that at time *t* an Archimedes-like scientist persuades (using very different, reliable methods) her colleagues that M is unreliable. This may be that society's first piece of scientific knowledge. The scientific community now rejects its earlier beliefs as unsound, realising that they were formed solely on the basis of a poor method.[20]

BE: On the semantic view this community was making progress until time *t* (it was accumulating true beliefs) and then regressed (it gave up those beliefs).[21]

BF: On the epistemic view that community made no progress at all until *t* at which time it did start to make progress.[22]

BG: In the history of science there are, for good reasons, no episodes that illustrate precisely such a divergence between truth and rationality.[23]

BH: Kuhn and Laudan do not think of solving a puzzle as involving knowledge, when knowledge is understood in the classical way as requiring truth.[24]

BI: Our conception of scientific progress is linked to what we take the aim of science to be. In general, something like the following principle holds: if the aim of X is Y, then X makes progress when X achieves Y or promotes the achievement of Y.[25]

BJ: If one is a skeptic, as Laudan is and Kuhn became, then the epistemic conception of progress will lead one to the conclusion that there has been no progress.[26]

---

[19] Bird [2007], p. 93.
[20] Bird [2007], p. 94.
[21] Bird [2007], p. 94.
[22] Bird [2007], p. 94.
[23] Bird [2007], p. 95.
[24] Bird [2007], p. 96.
[25] Bird [2007], p. 111.
[26] Bird [2007], p. 115.

BK: Laudan thinks that it is an advantage of his account that progress can be assessed internally. ... The semantic account makes progress more difficult and worthwhile, by relating it to the other clear benefits of truth.[27]

BL: In a community which encourages belief on flimsy evidence, scientific beliefs will come and go. On the semantic view of progress an episode in which a truth is believed by accident and then abandoned will count as progress followed by regress, whereas on the epistemic view, there will have been neither. This means that the view of science as having shown for the bulk of its history a continuous and monotonic growth of progress is easier to maintain on the epistemic than on the semantic view. This supports the contention that the thesis that science progresses (as conceived on the epistemic view) is the appropriate slogan for scientific results. Such an approach also avoids the problem of saying what exactly increasing verisimilitude amounts to for a large, diverse, and growing body of beliefs, taken all together.[28]

BM: Far from being internally accessible, like many of the best things in life, the most exciting contributions to progress are often recognizable as such only with the benefit of hindsight.[29]

Let's begin with statement BA. Since we have already made some progress in distinguishing eight main zones of inquiry, might we not ask 'What kind of *characterizing?*' Is it to be one of eight kinds of characterization normal to each of the eight main tasks?[30] Or, perhaps some other kind? Statement BB offers some answer to the question, as well as what the method shall be: "I shall argue ....[31]" The method employed is object-oriented, logical argument, in general philosophical terms, comparing logically presented views, terminologies and word usages in philosophical discussions about scientific progress. The specific aim of the paper is to argue for "a simple minded cumulative knowledge account of progress."[32] But, it is also to reach a logically defensible view of other views of scientific progress – past, present and possible. Note, too, that the argument given in Bird's article partly hinges on logic about

---

[27] Bird [2007], p. 115.

[28] Bird [2007], p. 115.

[29] Bird [2007], p. 115.

[30] Each of the eight main tasks will have its characters and characterizations. See Chapters 8 and 9.

[31] See note 18.

[32] See note 18.

supposed or imagined scientific communities. For all of this, see, for example, BD, BE, BF, BI, BJ, BL (part of Bird's concluding argument) and BM.

In as much as the work aims for a comprehensive view of all that has been done so far, the work leans toward (pre-) functional dialectics. But, there is no personal stand taken and, notably, no contact or verification in actual progress in science (such as, say, in the gradual emergence of new species of understanding in the hard-won dawning of gauge theory[33]). Bird argues in general terms that apply to an undefined range of imagined scientific communities. However, some descriptive results are stated as relatively permanent truths about progress in science. So, while not tied to instances of progress in science as such, the article's feet also cross over into (pre-) functional doctrines.

Statement BC regards two authors: "Popper, somewhat forlornly, hoped, ..." ; and "Ilkka Niiniluoto ... sought to build...." Statement BH purports to state what both Kuhn and Laudan each did *not* think, while BK asserts what "Lauden thinks ..." All of these claims are hermeneutical, a highly non-trivial task in contemporary scholarship. But, the statements are only fragments within the main fabric of the article-length argument. Justifications for the interpretative statements are not given. Hermeneutical principles employed are not indicated. Nor, as an alternative, is there appeal to the authority of exegetes in the field.

A.4   Discussion

A.4.1   Jammer: *Concepts of Mass*

As partially revealed in quotations above (and further evidenced by a more complete reading of the book), pages, sections and chapters of the book are a complex weave-work of threadings and fragments of distinct tasks. Tasks present in the book include hermeneutics; and tracking combinations of terminologies and arguments from history (but without discussion of developments).

---

[33] See Chapter 3, above, and O'Raifeartaigh's book referenced therein.

The dominant orientation is past-oriented:

> This book intends to provide a comprehensive and self-contained study of the
> concept of mass as defined, employed, and interpreted in contemporary theoretical
> and experimental physics and as critically examined in the modern philosophy of
> science. It studies in particular how far, if at all, present-day physics contributes to
> a more profound understanding of the nature of mass.[34]

Jammer's conclusion is stated in the last sentence of the Preface, and
again in the last paragraph of the book:

> Thus, in spite of all the strenuous efforts of physicists and philosophers, the notion
> of mass, although fundamental in physics, is, as we noted in the preface, still
> shrouded in mystery.[35]

There are two main features of the book to which I invite the reader's
attention: (1) The prevailing method is logical analysis of terminologies,
formulas and concepts; and (2) The book does not keep to a single task.
Logical analyses given do not ask that one advert to actual scientific
experience. Not attending to experience of doing science (neither
experimental, nor questions, nor insights nor choices) leads to various
difficulties, including, for example, apparent problems of "circular
logic."[36] Not keeping to any single task, the book does not build up to
focused progressive results.[37]

The problems I mention here, though, are not unique to Jammer's
work. In contemporary history and philosophy of physics, the two features
mentioned are normal, but generally not adverted to. Or, if they are

---

[34] Jammer [2000], p. vii.

[35] Jammer [2000], p. 167; and p. ix.

[36] "If 'force' is defined, as it generally is, as the product of acceleration and mass, then the
definition is obviously circular" (Jammer [2000], p. 6). "It is therefore interesting to note
that the very first proof of this relation ($E = mc^2$) – Einstein's 1905 derivation – has been
criticized as being a logical fallacy involving a vicious circle" (Jammer [2000], p. 62). See
also pp. 143ff.

[37] See Section A.1, Introduction. Contrast with, for example, the effectiveness of the
ongoing collaboration between experimental and theoretical physics. Well vetted results
from CERN, for instance, are handed on to theoreticians; and, going the other way,
questions and insights from theoreticians are shared with research teams working at, for
example, CERN.

noticed, they usually are not considered problematic. Note, too, that this has been only an extremely brief discussion about just a few extracts from a large book that was written by a senior scholar in the history and philosophy of physics. In fact, the book is a rich source of nuanced scholarship on the story of *mass* in 20th century physics and philosophy of physics. We need to also ask, therefore, how the book might contribute to progress of physics and philosophy of physics.

Might we not ask, for example, whether or not there might be anomalies in the tradition subtly revealed within (the dense weave of) the book? The scientific community struggles with a lack of control of meaning about *mass*, and other terms in contemporary theories. Was Jammer partly on to this? Perhaps Jammer's apparently *inconclusive conclusion*[38] was him touching on a core problem of present-day thought. Perhaps Jammer's work itself calls for interpretation? If that is so, might it not be helpful if key elements of his work were presented in ways that revealed possible divergence (positive or negative) from standard views? And, might such a contribution not help lighten the burden of those who choose to go on to hermeneutical investigations of Jammer's book and his life's work? Within (a future) explanatory heuristics, some might ask how aspects of Jammer's book might contribute (positively, or negatively) to progress. And so on. In other words, given the evident richness of the book, starting with (textual) research, and then (hermeneutical) interpretation, it probably will be important to recycle the text up and into (in the not too distant future, some approximation to) an eightfold functional collaboration.

A.4.2   Bird: *Scientific Progress*

The dominant mode of the paper is logical argument about wordings and terminologies in various "conceptions"[39] of progress. Discussions do not appeal to experience in science, but speak in general terms about, for example, *concepts, theory, knowledge, truth, meaning* and *progress*.

---

[38] See note 35.
[39] See, for example, titles of sections of the article.

As indicated at the beginning of Bird's paper (BB), there is a leaning toward (pre-) functional dialectics: "I will argue ..." However, logical arguments also depend on supposed aspects of imagined scientific communities. Bird argues for truths about all possible progress in science, and in that way his work also has its feet partly in (pre-functional) doctrines.

Throughout the paper, there are hermeneutical claims. See, for example, BC, BH, BJ and BK. These quotations are only a small sampling of numerous similar statements throughout the article, statements that claim what authors mean, think or do not think, and hope. These claims mainly regard the terminologies and wordings of authors. The interpretative statements are fragments within the main logical arguments of the paper. In particular, hermeneutical claims appeal neither to sources nor to the works of exegetes.

While there are claims about the meanings of authors across several decades, thus revealing an historical dimension to the paper, there is no discussion of developments in the tradition. Instead, alleged meanings of authors are referred to as elements in a discrete set.

There is ongoing transitioning among different tasks, often occurring within single paragraphs. Logical arguments about hypothetical or merely nominally defined terms are remote to actual scientific progress. The ongoing transitioning from one task to another, and in some cases mergings of tasks, undermines the possibility of contributing significantly to any of the eight main tasks.

As in the discussion about Jammer's work, here too, these problems are not unique to Bird's results, but are representative of scholarship in contemporary philosophy of science. And so, here too, we can ask how Bird's work might contribute to progress. One key question regards method: Is the method of logical argument employed by Bird a sufficient basis for

comparing ... approaches to characterizing scientific progress[40]?

---

[40] See note 17.

Among other things, asking the question reveals that there is more to characterizing progress than terminologies and conceptions, and logical arguments about terminologies and conceptions. Exploring anomalies here could eventually lead to a positive recycling of Bird's scholarship, and, indeed, other works in the same tradition.

A.5 References

Bird, A. (2007) What is Scientific Progress? *Noûs*, vol. 41, issue 1, pp. 92–117.

Jammer, M. (2000) *Concepts of Mass in Contemporary Physics and Philosophy* (Princeton University Press, Princeton, NJ).

Index

Printed in the United States
By Bookmasters